BURLEIGH DODDS SCIENCE: INSTANT INSIGHTS

NUMBER 113

Organic soil amendments

Published by Burleigh Dodds Science Publishing Limited
82 High Street, Sawston, Cambridge CB22 3HJ, UK
www.bdspublishing.com

Burleigh Dodds Science Publishing, 1518 Walnut Street, Suite 900, Philadelphia, PA 19102-3406, USA

First published 2025 by Burleigh Dodds Science Publishing Limited
© Burleigh Dodds Science Publishing, 2025. All rights reserved.

British Library Cataloguing in Publication Data
A catalogue record for this book is available from the British Library

ISBN 978-1-83545-248-6 (Print)
ISBN 978-1-83545-249-3 (ePub)

DOI: 10.19103/9781835452493

Typeset by Deanta Global Publishing Services, Dublin, Ireland

Contents

Acknowledgements   v

1   Organic fertilizers and biofertilizers   1
    Lidia Sas Paszt and Slawomir Gluszek, Research Institute of
    Horticulture, Poland

    1 Introduction   1
    2 Biofertilizers   2
    3 Consortia of microorganisms to improve the effectiveness of organic
      fertilization   7
    4 Animal excrement: manures, slurry and guano   9
    5 Products and by-products of animal origin   10
    6 Products and by-products of plant origin for fertilizers   12
    7 Composts   19
    8 Untreated minerals and by-products of selected industrial processes   21
    9 Biochar   21
    10 Conclusion   22
    11 Where to look for further information   23
    12 References   23

2   Assessing the effects of compost on soil health   41
    Cristina Lazcano, University of California-Davis, USA;
    Charlotte Decock, California Polytechnic State University, USA;
    Connie T. F. Wong, University of California-Davis, USA; and
    Kamille Garcia-Brucher, California Polytechnic State University, USA

    1 Introduction   41
    2 Why compost?   44
    3 Effects of compost on soil nutrient cycling   47
    4 Effects of compost on soil hydraulic properties   53
    5 Effect of compost on crop productivity   55
    6 Effects of compost on soil biodiversity   59
    7 Effects of compost on environmental quality   63

8 The use of compost to improve soil health in annual crops: a case study
with strawberries   66

9 The use of compost to improve soil health, sequester carbon and
reduce greenhouse gas emissions in perennial crops: a case study in a
Mediterranean vineyard   72

10 Conclusion   78

11 Where to look for further information   79

12 References   80

3   Optimizing slurry management   101
*David Fangueiro, LEAF-Instituto Superior de Agronomia-ULisboa,
Portugal; Jihane Elmahdi, Wageningen University and Research,
The Netherlands; Jared Nyang'au, Aarhus University, Denmark;
Stamatis Chrysanthopoulos, LEAF-Instituto Superior de Agronomia-
ULisboa, Portugal; Jerke De Vries, Wageningen University and
Research, The Netherlands; and Peter Sørensen, Aarhus University,
Denmark*

1 Introduction   101

2 Current decision tools for optimizing manure management   103

3 Modifying animal slurry pH to enhance its value as a biobased fertilizer:
(bio)-acidification and alkalinization   107

4 Improving manure management systems to minimize trade-offs   112

5 Combining manure management with anaerobic digestion   117

6 Pre- and post-treatment of biomass for anaerobic digestion   120

7 Optimization of anaerobic digestion operations to optimize digestate quality 124

8 References   125

4   Optimizing livestock manure as a biofertilizer and bioenergy source   135
*V. Riau, L. Morey, R. Cáceres, M. Cerrillo and A. Bonmatí, Institute
of Agrifood Research and Technology (IRTA), Spain; and A. Robles,
BETA Tech Center (UVIC-UCC), Spain*

1 Introduction   135

2 Anaerobic digestion   138

3 Mechanical separation   141

4 Composting   145

5 Struvite precipitation   149

6 Stripping/scrubbing   152

7 Membrane filtration   154

8 Bioelectrochemical systems   158

9 Case study: farm for the future   160

10 Conclusion and future trends   165

11 Where to look for further information   165

12 References   166

Series list

Title	Series number
Sweetpotato	01
Fusarium in cereals	02
Vertical farming in horticulture	03
Nutraceuticals in fruit and vegetables	04
Climate change, insect pests and invasive species	05
Metabolic disorders in dairy cattle	06
Mastitis in dairy cattle	07
Heat stress in dairy cattle	08
African swine fever	09
Pesticide residues in agriculture	10
Fruit losses and waste	11
Improving crop nutrient use efficiency	12
Antibiotics in poultry production	13
Bone health in poultry	14
Feather-pecking in poultry	15
Environmental impact of livestock production	16
Sensor technologies in livestock monitoring	17
Improving piglet welfare	18
Crop biofortification	19
Crop rotations	20
Cover crops	21
Plant growth-promoting rhizobacteria	22
Arbuscular mycorrhizal fungi	23
Nematode pests in agriculture	24
Drought-resistant crops	25
Advances in detecting and forecasting crop pests and diseases	26
Mycotoxin detection and control	27
Mite pests in agriculture	28
Supporting cereal production in sub-Saharan Africa	29
Lameness in dairy cattle	30
Infertility and other reproductive disorders in dairy cattle	31
Alternatives to antibiotics in pig production	32
Integrated crop–livestock systems	33
Genetic modification of crops	34

Developing forestry products	35
Reducing antibiotic use in dairy production	36
Improving crop weed management	37
Improving crop disease management	38
Crops as livestock feed	39
Decision support systems in agriculture	40
Fertiliser use in agriculture	41
Life cycle assessment (LCA) of crops	42
Pre- and probiotics in poultry production	43
Poultry housing systems	44
Ensuring animal welfare during transport and slaughter	45
Conservation tillage in agriculture	46
Tropical forestry	47
Soil health indicators	48
Improving water management in crop cultivation	49
Fungal diseases of apples	50
Using crops as biofuel	51
Septoria tritici blotch in cereals	52
Biodiversity management practices	53
Soil erosion	54
Integrated weed management in cereal cultivation	55
Sustainable forest management	56
Restoring degraded forests	57
Developing immunity in pigs	58
Bacterial diseases affecting pigs	59
Viral diseases affecting pigs	60
Developing immunity in poultry	61
Managing bacterial diseases of poultry	62
Proximal sensors in agriculture	63
Dietary supplements in dairy cattle nutrition	64
Dietary supplements in poultry nutrition	65
Intercropping	66
Managing arthropod pests in tree fruit	67
Weed management in regenerative agriculture	68
Integrated pest management in cereal cultivation	69
Economics of key agricultural practices	70
Nutritional benefits of milk	71

Biostimulant applications in agriculture   72

Phosphorus uptake and use in crops   73

Optimising pig nutrition   74

Nutritional and health benefits of beverage crops   75

Artificial Intelligence applications in agriculture   76

Ensuring the welfare of laying hens   77

Ecosystem services delivered by forests   78

Improving biosecurity in livestock production   79

Rust diseases of cereals   80

Optimising rootstock health   81

Irrigation management in horticultural production   82

Improving the welfare of gilts and sows   83

Improving the shelf life of horticultural produce   84

Improving the health and welfare of heifers and calves   85

Managing arthropod pests in cereals   86

Optimising photosynthesis in crops   87

Optimising quality attributes in poultry products   88

Advances in fertilisers and fertiliser technology   89

Sustainable tropical forest management   90

Phenotyping applications in agriculture   91

Fungicide resistance in cereals   92

Unmanned aircraft systems in agriculture   93

Using manure in soil management   94

Zero/no till cultivation   95

Optimising agri-food supply chains   96

Optimising quality attributes in horticultural products   97

Improving the sustainability of dairy production   98

Improving the welfare of growing and finishing pigs   99

Ensuring the welfare of broilers   100

Biofertiliser use in agriculture   101

Infertility and other reproductive disorders in pigs   102

Regenerative techniques to improve soil health   103

Agroforestry practices   104

Carbon monitoring and management in forests   105

Viruses affecting horticultural crops   106

Regulatory frameworks for new agricultural products and technologies   107

Machine vision applications in agriculture   108

Alternative sources of protein for poultry   109

Alternative sources of protein for pigs   110

Nitrogen-use efficiency   111

Good agricultural practices (GAP)   112

Organic soil amendments   113

Developing a circular economy   114

Economics of agriculture   115

Controlled environment agriculture   116

Understanding and conserving pollinators   117

Land use change and management   118

Fusarium wilt of banana   119

Novel biocontrol agents   120

Soil carbon sequestration   121

Soil contaminants   122

Environmental impact of poultry production   123

Silvopastoral systems   124

Acknowledgements

Chapters in this Instant Insight are taken from the following sources:

Chapter 1 Organic fertilizers and biofertilizers
   Chapter taken from: Köpke, U. (ed.), Improving organic crop cultivation, Burleigh Dodds Science Publishing, Cambridge, UK, 2019, (ISBN: 978 1 78676 184 2)

Chapter 2 Assessing the effects of compost on soil health
   Chapter taken from: Horwath, W. R. (ed.), Improving soil health, Burleigh Dodds Science Publishing, Cambridge, UK, 2023, (ISBN: 978 1 78676 670 0)

Chapter 3 Optimizing slurry management
   Chapter taken from: Chapter taken from: Amon, B. (ed.), Developing circular agricultural production systems, Burleigh Dodds Science Publishing, Cambridge, UK, 2024, (ISBN: 978 1 80146 256 3)

Chapter 4 Optimizing livestock manure as a biofertilizer and bioenergy source
   Chapter taken from: Amon, B. (ed.), Developing circular agricultural production systems, Burleigh Dodds Science Publishing, Cambridge, UK, 2024, (ISBN: 978 1 80146 256 3)

Chapter 1

Organic fertilizers and biofertilizers

Lidia Sas Paszt and Slawomir Gluszek, Research Institute of Horticulture, Poland

1 Introduction

2 Biofertilizers

3 Consortia of microorganisms to improve the effectiveness of organic fertilization

4 Animal excrement: manures, slurry and guano

5 Products and by-products of animal origin

6 Products and by-products of plant origin for fertilizers

7 Composts

8 Untreated minerals and by-products of selected industrial processes

9 Biochar

10 Conclusion

11 Where to look for further information

12 References

1 Introduction

Organic fertilizers contain organic substances, including amino acids, sugars, lipids, vitamins, enzymes, plant hormones, macro- and microelements and other active compounds, which stimulate plant growth and development (Crouch and van Staden, 1993; Stirk and Staden, 1996; Khan et al., 2009; Ji et al., 2017).

Products permitted for use in organic farming in the European Union (EU) are listed in 'Annex I' to the EEC Regulation 2092/1991. At present, there is 'Annex II' – 'Products for use in fertilization and soil conditioning' (Løes et al., 2016). This list covers a wide range of products, from manures and composts to sediments from freshwater bodies. All of them are used as raw materials or, nowadays, as highly processed, modern products in the form of concentrated liquids, enriched powders, granulates or similar products. New bio-based formulations make it possible to save energy, labour input and time needed for efficient field application and plant nutrition (Vaneeckhaute et al., 2013). Biofertilizers contain, among other components, living cells of microorganisms

http://dx.doi.org/10.19103/AS.2017.0029.08

as the biologically active ingredient. The products suitable for organic agriculture that are available on the European market include a large group of products ranging from dried manures to highly processed organic-mineral fertilizers and mycorrhizal and bacterial inocula. Despite numerous studies on the use of preparations of organic origin, there is still a lack of bio-stimulants and protective substances for enriching the habitats of plants microbiologically. Further work is necessary on improving the technologies of cultivation and fertilization of fruit plants.

2 Biofertilizers

There has been, so far, no precise definition for the term 'biofertilizer'. Some authors define it as a substance that contains, as an active ingredient, live cells of microorganisms separately or in combination with other active ingredients that colonize the rhizosphere or spaces inside plant tissues and that help in enhancing the nutrient uptake by plants, and is thus able to enrich the nutrient quality of the soil (Vessey, 2003). This definition limits biofertilizers to microbes that can assimilate free nitrogen or dissolve water-insoluble salts of phosphorus, potassium or other nutrients. Sometimes, the term 'biofertilizer' is used for artificially multiplied microbial inoculants that can improve soil fertility and crop productivity (Roychowdhury et al., 2015).

Others define biofertilizers as biologically active products containing organic compounds such as amino acids, sugars, vitamins and other substances that have a direct nutritional impact on plant growth and yielding.

However, limiting the action only to nutritional effects means that in many cases, there is a problem with a wide group of organic products (both microbial and of organic origin) which also have other effects, for example, protection against diseases or pests or plant growth regulation.

A large group of compounds for agriculture, especially for organic farming, that contain living microbial cells are called biofertilizers. This category covers products that contain live cells of bacteria, cyanobacteria, actinomycetes, yeast cells and spores and/or hyphae of microscopic, filamentous or mycorrhizal fungi. The bacterial and other microbial strains used in the products for organic agriculture belong to various genera and species.

The mechanisms by which plant-growth-promoting rhizobacteria (PGPR) enhance plant growth are not fully understood. However, it is believed that the PGPR promote plant growth and yielding by either direct or indirect mechanisms, where direct mechanisms include the ability to produce phytohormones like indole acetic acid (IAA), gibberellin, cytokinins and ethylene; asymbiotic N_2 fixation; antagonism against phytopathogenic microorganisms by the production of siderophores; and solubilization of mineral phosphates and other nutrients; and indirect mechanisms entail the extracellular production of antibiotics, synthesis of antifungal metabolites, production of fungal cell wall lysin enzymes, depletion of iron from the rhizosphere and competition for sites on roots and induced systemic resistance (Salantur et al., 2006; Egamberdiyeva, 2007; Ahmad et al., 2008; Aznar et al., 2014; Gusain et al., 2015).

2.1 Nitrogen-fixing microorganisms

'Traditional' biofertilizers contain symbiotic bacteria, for example, rhizobia (*Rhizobium, Neorhizobium, Allorhizobium, Pararhizobium, Mesorhizobium, Bradyrhizobium* and *Ensifer*), free-living bacteria, such as *Azotobacter, Azospirillum,* or cyanobacteria like *Nostoc, Anabaena, Tolypothrix* or *Aulosira,* which are able to fix free nitrogen from the air (Souza et al., 2014; Dhar et al., 2015).

2.1.1 Rhizobia

Rhizobia are bacteria that formerly belonged to the family Rhizobiaceae, but at present, some of them are classified into new families: Phyllobacteriaceae into the genus *Mesorhizobium,* Nitrobacteriaceae (Bradyrhizobiaceae) into the genus *Bradyrhizobium* and Burkholderiaceae into the genus *Paraburkholderia* (formerly *Burkholderia*) (Velázquez et al., 2017). The complete list of valid species of rhizobia is constantly updated and recorded as the 'List of Prokaryotic names with Standing in Nomenclature' (http://www.bacterio.cict.fr).

These bacteria are mutualistic symbionts, and fix nitrogen environmentally depended highly variable at about 50–350 kg/ha, with legumes only. They are useful for pulse legumes like chickpea, red-gram, pea, lentil, black gram and so forth; oilseed legumes like soybean and groundnut; and forage legumes like berseem, clovers and lucerne. Successful nodulation of leguminous crops by a *Rhizobium* largely depends on the availability of compatible strains for a particular legume. Rhizobia colonize the roots of specific legumes to form tumour-like growths called root nodules, which act as factories of ammonia production. Rhizobia have the ability to fix atmospheric nitrogen in a symbiotic association with legumes and certain non-legumes, like *Parasponia*. Each legume requires a specific species of *Rhizobium* to form effective nodules. Many legumes may be nodulated by diverse strains of Rhizobia, but growth is enhanced only when nodules are produced by the effective strains of Rhizobia. It is thus extremely important to match microsymbionts prudently for maximum nitrogen fixation, especially because some native strains are not as effective as host-specific rhizobial strains, and there is a possibility of establishing symbiosis with wild 'cheating', non-effective nitrogen-fixing strains (Kiers et al., 2003; Vicario et al., 2016). A strain of *Rhizobium* that nodulates and fixes a large amount of nitrogen in association with one legume species may also do the same in association with certain other legume species. Leguminous plants demonstrate the tendency to respond similarly to particular strains of rhizobia (Wani and Lee, 2002).

The population of *Rhizobium* in the soil depends on the presence of legume crops in the field. In the absence of legumes, its population decreases. Artificial seed inoculation is often needed to restore the population of effective *Rhizobium* strains near the rhizosphere to hasten N_2-fixation. Symbiotic associations of legumes and rhizobia can be a major source of nitrogen in organic and sustainable systems, whereas free-living nitrogen-fixing microorganisms are helpful in non-symbiotic systems (Rashid et al., 2016a).

2.1.2 Azospirillum

Azospirillum belongs to the family Spirillaceae and is heterotrophic and associative in nature. It has nitrogen-fixing ability of about 20-40 kg/ha. Besides nitrogen fixation, bacteria from the genus *Azospirillum* are also able to produce phytohormones: auxins, gibberellins, cytokinins and phytohormone-like agents, such as indole derivatives (Cassán et al., 2014). Although there are many species under this genus, such as *Azospirillum amazonense*, *Azospirillum halopraeferens* and *Azospirillum brasilense*, the worldwide distribution and benefits of inoculation have been proved mainly with the *Azospirillum lipoferum* and *Azospirillum brasilense*. *Azospirillum* forms associative symbiosis with many plants, particularly with those having the C4-dicarboxylic pathway of photosynthesis (the Hatch and Slack pathway), because they grow and fix nitrogen on salts of organic acids such as malic and aspartic (Arun, 2007). Thus, they are mainly recommended for maize, sugarcane, sorghum, pearl millet and so forth, especially in environmental stress conditions like drought (Moutia et al., 2010). *Azospirillum* colonizing the roots do not only remain on the root surface, but a sizable proportion of them also penetrate into the root tissues and live in harmony with the plants. They do not, however, produce any visible nodules or outgrowth on root tissue.

2.1.3 Azotobacter

Azotobacter belongs to the family Azotobacteraceae, and is aerobic, free living and heterotrophic in nature. *Azotobacter* are present in neutral or alkaline soils, with *Azotobacter chroococcum* being the most commonly occurring species in arable soils. *Azotobacter vinelandii*, *Azotobacter beijerinckii*, *Azotobacter insignis* and *Azotobacter macrocytogenes* are other reported species. The number of *Azotobacter* rarely exceeds 10^4-10^5 g^{-1} of soil due to lack of organic matter and presence of antagonistic microorganisms in the soil. The bacterium produces antifungal antibiotics which inhibit the growth of several pathogenic fungi in the root region, thereby preventing seedling mortality to a certain extent (Nagaraja et al., 2016). An isolated culture of *Azotobacter* fixes about 10 mg nitrogen g^{-1} of carbon source under *in vitro* conditions. *Azotobacter* are also known to synthesize biologically active growth-promoting substances such as vitamins of the B group, IAA and gibberellins. Many strains of *Azotobacter* also exhibit antifungal properties against plant pathogens such as *Fusarium* or *Alternaria* (Chetverikov and Loginov, 2009). They can also be applied in environmental stress management (Di Stasio et al., 2017). The population of *Azotobacter* is generally low in the rhizosphere of crop plants and in uncultivated soils. The occurrence of this organism has been reported in the rhizosphere of a number of crop plants such as rice, maize, sugarcane, vegetables and plantation crops (Arun, 2007).

2.1.4 Blue-green algae (Cyanobacteria)

Blue-green algae belong to eight different families and are phototrophic in nature, producing auxin, IAA and gibberellic acid, fixing about 20–30 kg N/ha in submerged rice fields. Since they are abundantly found in paddies, they are also referred to as 'paddy organisms'. N is the key input required in large quantities for lowland rice production. Soil N and biological nitrogen fixation (BNF) by associated organisms are major sources of N for lowland rice. The 50–60% N requirement is met through the combination of mineralization of soil organic N and BNF by free-living and rice-plant-associated bacteria (Roger and Ladha, 1992). To achieve food security through sustainable agriculture, the requirement for fixed nitrogen must be increasingly met by BNF rather than by industrial nitrogen fixation. Most N_2-fixing BGA are filamentous, consisting of a chain of vegetative cells, including specialized cells called heterocysts which function as micronodules for the synthesis and N_2-fixing process. BGA forms a symbiotic association capable of fixing nitrogen with fungi, liverworts, ferns and flowering plants. The most effective common symbiotic association has been found for the *Azolla* and *Anabaena azollae* (BGA) complex. Besides N_2 fixation, BGA are also capable of producing siderophores which can complex metal ions into stable forms (Singh et al., 2016).

   Strains of other free-living and endophytic bacterial species are also capable of fixing nitrogen from the air (Fox et al., 2016). Nitrogen-fixing microorganisms seem to be an economically attractive and ecologically friendly alternative to artificial nitrogen fertilizers, especially in sustainable agricultural production.

2.2 Non-nitrogen-fixing bacteria

Non-nitrogen-fixing bacteria are also suitable for agricultural use both as biofertilizers and plant growth promoters, especially as nutrient (e.g. phosphate) solubilizers, which are able to solubilize inorganic phosphate compounds, such as tricalcium phosphate, dicalcium phosphate, hydroxyapatite and rock phosphate. Among the bacterial genera with this capacity are *Pseudomonas, Bacillus, Rhizobium, Burkholderia, Achromobacter, Agrobacterium, Micrococcus, Aerobacter, Flavobacterium* and *Erwinia*. There are considerable populations of phosphate-solubilizing bacteria in the soil and the plant rhizosphere. These include both aerobic and anaerobic strains, with a prevalence of aerobic strains in submerged soils. A considerably higher concentration of phosphate-solubilizing bacteria is commonly found in the rhizosphere in comparison with the bulk soil (Roychowdhury et al., 2015). The soil bacteria which belong to the genera *Pseudomonas* and *Bacillus*, and fungi are more common. The major microbiological means by which insoluble P compounds are mobilized is by the production of organic acids, accompanied by acidification of the micro-environment (Raghu and MacRae, 1966). The organic and inorganic acids convert tricalcium phosphate to di- and monobasic phosphates with the net result of an enhanced availability of the element to the plant. The type of organic acids produced and their amounts differ with different organisms (Chen et al., 2006). Tri- and di-carboxylic acids are more effective compared to monobasic and aromatic acids. Aliphatic acids are also found to be more effective in P solubilization compared to phenolic, citric and fumaric acids (Walpola and Yoon, 2012). The analysis of culture filtrates of phosphate-solubilizing microorganisms has shown the presence of a number of organic acids, including citric, fumaric, lactic, 2-ketogluconic, gluconic, glyoxylic and ketobutyric acids (Wani et al., 2015).

Other rhizosphere microorganisms can reduce manganese and iron into water-soluble forms which are available for uptake by plants. In particular, bacteria belonging to the genera *Bacillus, Pseudomonas, Citrobacter, Shewanella, Alcaligenes, Clostridium* and *Enterobacter* are capable of reducing insoluble salts of iron and manganese (Ebrahiminezhad et al., 2017). The reduction of metal oxides in the soil can be linked to anaerobic oxidation of methane, one of the important greenhouse gases in low-oxygen soil conditions (Oni and Friedrich, 2017). Some beneficial bacteria, like *Pseudomonas, Rhizobium, Azotobacter* or *Erwinia* sp., can produce chelating agents (siderophores) that can form soluble complexes with soluble ions, preventing them from reprecipitation (Born et al., 2016; Kumar et al., 2017). The production of chelating agents is also one of the mechanisms of pathogen suppression by beneficial microorganisms (Sasirekha and Srividya, 2016).

2.3 Mycorrhiza

Another important group of beneficial microorganisms are mycorrhizal fungi. In agriculture and horticulture, the most common are arbuscular mycorrhizal fungi (AMF) belonging to Glomeromycota. They are obligatory symbionts of plants and form symbiosis with almost 80% of plant species on the earth (Davison et al., 2015; Öpik et al., 2010). These fungi are associated with the majority of agricultural crops, except the crops/plants belonging to the families of Chenopodiaceae, Amaranthaceae, Caryophyllaceae, Polygonaceae, Brassicaceae, Commelinaceae, Juncaceae and Cyperaceae. AMF can produce a network of extraradical mycelia, which gives access to a greater quantity of water and soil minerals not available to the roots of the host plants. AMF also produce the peptide-like substance – glomalin, which plays a key role in the stabilization of soil structure and quality (Singh et al., 2013; Bedini et al., 2010; Vasconcellos et al., 2016). The fungi receive the necessary carbohydrates from plants for the completion of their life cycle. Arbuscular mycorrhizal associations have been shown to reduce damage caused by soil-borne plant pathogens. It is well documented that the arbuscular mycorrhizal symbiosis can increase plant growth and nutrient uptake, improve fruit quality and alleviate several abiotic stresses such as low temperature stress, drought, salt stress and so forth (Cavagnaro et al., 2015; Rouphael et al., 2015; Taktek et al., 2015; Zhao et al., 2015; Yooyongwech et al., 2016).

Application of microorganisms can increase the bioavailability of nutrients and the stability of soils and can support formation of soil aggregates in poor soils (Rashid et al., 2016b).

3 Consortia of microorganisms to improve the effectiveness of organic fertilization

Application of selected microorganisms may increase the effectiveness of organic fertilizers in plant production. Especially effective is co-application of microorganisms with organic fertilizers.

Microorganisms can be applied alone, as a single strain, but in many cases, better effects are observed after application of microbial consortia. The increased effects are achieved through the synergic relation between different groups of microorganisms. Application of consortia increases the nutritional effect of organic fertilizers, with additional 'value added' contributions, such as the production of phytohormones, antimicrobial agents, siderophores and others. Pseudomonads applied with AMF stimulate flower production and increase the yielding of strawberry, increasing yield parameters and amounts of biologically active substances in strawberry fruits (Bona et al., 2015). Microbial consortia are helpful in better utilization of organic or mineral fertilizers.

Grzyb et al. (2015) have shown that the application of the commercially available microbial consortium 'Mycosat' with different organic preparations and fertilizers increased plant growth in sour cherry 'Debreceni Bötermö' and apple 'Topaz' than the application of fertilizers alone. 'Mycosat' stimulated the growth of plants and promoted better bud-take on inoculated rootstocks. The resistance of bud grafts to freezing was increased. A disadvantage, however, is its relatively low effectiveness when the soil has become excessively dry. A double application of this preparation during the growing season gave better results in the nursery than a one-off treatment.

Common applications of AMF (*Rhizophagus intraradices*, *Glomus aggregatum*, *Glomus viscosum*, *Claroideoglomus etunicatum* and *Claroideoglomus claroideum*) and plant-growth-promoting *Pseudomonas* strains (C7 and 19Fv1T) positively affected the flowering and fruit production, and also the amounts of vitamin C, sugars and organic acids in tomato plants cv. 'TC 2000' (Bona et al., 2017).

Application of a mixture of beneficial bacteria namely *Pantoea* sp., *Pseudomonas fluorescens*, *Klebsiella oxytoca* and *Rhizobium* sp. on organically or mineral-NPK fertilized fields caused better plant growth and yielding of apple cv. 'Topaz' in comparison with application of single mineral or organic fertilizers alone (Mosa et al., 2016). Preparations with AMF phosphate-solubilizing bacteria and *Azotobacter* combined with vermicompost increased the growth and yielding of strawberry plants in comparison with vermicompost single application (Singh et al., 2015). Similar effects were observed in basil (*Ocimum basilicum*) after co-application of vermicompost, *Dietzia natronolimnaea* and *Glomus intraradices* (Bharti et al., 2016b).

The synergic effect of different microorganisms is taken advantage of in the production of commercial products which contain beneficial bacterial strains and arbuscular or filamentous fungi, sometimes mixed with rock meal as a source of phosphorus, potassium and other minerals (Grzyb et al., 2015).

A group of products called 'Effective Microorganisms' (EM), popular in organic agriculture, is obtained during molasses fermentation (see Section 6.1) using a consortium of many microbial strains. EM, when applied as fertilizer, resulted in positive impacts on plant growth parameters (Kleiber et al., 2014; Salama et al., 2014). For example, strawberry plants cv. 'Honeoye' treated with EM gave similar results to those under mineral fertilization, whereas cv. 'Selva' showed better growth parameters after the application of 'Effective Microorganisms' (Glinicki et al., 2011). Indirect, positive effects in plant growth and yielding are obtained when 'Effective Microorganisms' are used as microbial inoculum for compost preparation. For example, in an eleven-year long-term experiment, wheat fertilized with 'Effective Microorganisms' compost gave better plant growth parameters and yield in comparison to traditional compost (Hu and Qi, 2013). 'Effective Microorganisms' are especially useful in co-application with organic matter like agricultural wastes or with bioorganic fertilizers while increasing the effectiveness of nutrients utilization (Chantal et al., 2013; Shaheen et al., 2017).

4 Animal excrement: manures, slurry and guano

The most common organic fertilizer used in organic farming is farmyard animal manure, which is the basis of organic fertilization of many crops. In recent years, there has been a problem with adequate amounts of this resource for both organic and integrated/conventional agriculture. Manure can be applied in fresh, dried or composted form. However, for safe organic production, to avoid the risk of contamination with pathogenic or opportunistic strains of microorganisms, it is recommended that manure be thermally processed, for example, during thermophilic composting. If raw manure is used, it should be applied 90 days before harvest for crops that do not come into direct contact with the soil, and 120 days prior to harvest for crops that have direct contact with the soil (Köpke et al., 2007). The use of manure originating from factory farming is forbidden.

Long-term application of farmyard manure increases the levels of P, K and S available to plants, and also the amounts of microelements like B, Zn, Fe or Mo in the soil, but can decrease the concentration of Cu and Mn (Rutkowska et al., 2014). Application of manures modifies the amounts of available forms of nutrients in the soil profile. This effect is dependent on the origin and form of the applied manure (Schlegel et al., 2017).

Manures are a source of soluble substances which are easily utilized by soil microorganisms. These substances are sugars, alcohols and volatile fatty acids such as acetic, butyric and valeric acids. Slowly biodegradable organic polymers must be hydrolysed before their utilization by soil microorganisms (Boursier et al., 2005). At present, manures are available on the market in processed form, dried or composted or as dried granules or pellets. Some of these products are supplemented with rock powders, animal blood, bone or feather meals, plant by-products or other materials (Malusà et al., 2014).

Other traditional products of animal origin, widely used as fertilizers, are slurry (liquid animal excrement) used after controlled fermentation and/or appropriate dilution.

Guano, accumulated excrement of seabirds, seals or cave-dwelling bats, is a highly effective fertilizer with high nutrient contents essential for plant growth. However, its mineral content is strictly associated with the feeding regimes of guano-producing animals (Szpak et al., 2012). For example, the guano produced by the sanguivorous bat *Desmodus rotundus* has more than 17% nitrogen, that produced by the frugivorous bat *Pteropus rodricensis* has about 2% and the guano produced by the insectivorous bat *Tadarida brasiliensis* has a nitrogen content of about 10% (Emerson and Roark, 2007). Large deposits of guano are formed in dry climates, where they are protected from nitrogen leaching by rain.

5 Products and by-products of animal origin

5.1 Raw or processed animal residues

A large group of organic fertilizers are raw or processed animal residues, dried and pulverized blood, hoof, horn, bone, degelatinized bone, fish, meat, feather, hair, 'chiquette' meal,'chiquette', wool, fur, hair, dairy products and hydrolysed proteins.

Animal residues are sources of organic matter and minerals N, P, Ca and others. But the usage of these materials, especially meat and bones, for agriculture is restricted by safety rules, particularly those relating to TSE (transmissible spongiform encephalopathy) incidents (Möller, 2015). Substrates allowed for the preparation of organic fertilizers without pressure sterilization are those classified to Category 3 according to the EC Regulation 1069/2009. By-products in this category are obtained from the processing of goods intended for human consumption obtained by slaughtering animals, leftovers from food processing facilities or canteens. This category of by-products, when used for the preparation of organic fertilizers, must be treated by pasteurization at 70°C for more than 1 h. Category 2 covers by-products from animals that died in ways other than slaughtering, animals killed to prevent epizootic diseases or those posing a risk of infection with diseases other than TSE. This category also covers the contents of digestive tracts obtained by slaughtering healthy animals. These residues may be used for the production of fertilizers after pressure sterilization at a temperature of more than 133°C for more than 20 min under a pressure above three bars. Processing of meals into fertilizers may be done by simple granulation to obtain a solid phase. Category 2 by-products can be processed by acid hydrolysis to obtain short peptide chains suitable for the production of solid nitrogen-rich fertilizers. For example, 'Biollsa', prepared by Ilsa Group, Italy, is a granular nitrogen fertilizer produced by acid hydrolysis.

Another group of fertilizers of animal origin, popular in recent years, are liquid hydrolysates rich in amino acids and short peptide chains. The best of these preparations are obtained by enzymatic hydrolysis of purified animal residues such as collagen. For maximum safety, some producers use only Category 3 residues to avoid the risk of spreading disease. Liquid formulation makes them easy to use by spraying, especially in pomological orchards, berries plantations, vineyards and fields. Liquid hydrolysates are also used via irrigation systems, in greenhouse production of high-quality vegetables and fruits. Hydrolysed proteins must not be applied to edible parts of the crop, and in the case of pastures, the feeding of animals must be prohibited for at least three weeks to avoid the potential risk of disease infection. Amino acids are also a source material for the production of complexes with micronutrients, which are eco-friendly plant growth promoters (Ghasemi et al., 2013). The effectiveness of amino acid preparations in plant nutrition and for improving soil quality has been investigated in many trials, which have shown that applications of such preparations increase the yield quantity and quality (Lisiecka et al., 2011; Corte et al., 2014; Colla et al., 2015; Mobini et al., 2014). For example, the Terra-Sorb Foliar amino acid preparation applied on oilseed rape increases the levels of fats and proteins and decreases the amounts of glucosinolates in seeds in comparison with the control (Jakienė, 2013).

Wool, fur and hair are a source of slowly released nitrogen when used in a low-processed form like pellets or meal. They are also used as a feedstock for the production of more processed preparations such as hydrolysates (Nustorova et al., 2006).

Dairy products and by-products, especially whey, have been a point of interest as a form of nitrogenous fertilizer for a long time (Sharratt et al., 1959; Peterson et al., 1979; Marwaha and Kennedy, 1988). Direct application of whey on arable fields could be a method for improving soil fertility and for improving the physical properties of some types of degraded soils (Robbins and Lehrsch, 1992; Jones et al., 1993).

Good results of whey utilization are obtained when it is used as a raw material for organic fertilizer production. Whey-based fertilizers are compositions of different organic amendments like sawdust and straw, and also with minerals like zeolite. An example of such a product is 'Condit' – a pelletized sawdust, straw and zeolite mixture enriched with whey as a source of nitrogen (Cuijpers and Hospers-Brands, 2008). Whey is also a good supplement for bioconversion of some substrates of organic origin into valuable organic fertilizers that have positive effects on plant growth parameters (Stępień et al., 2012). They are also an effective amendment for compost teas, especially in the protection against plant pathogens (Pane et al., 2012).

5.2 Chitin

Chitin fertilizers are obtained from shells of crustaceans. This product is allowed only if obtained from sustainable fisheries, as defined in EC 2371/2002, or organic aquaculture. Chitin and its derivatives are not only sources of nitrogen and carbon, but also active against viruses, bacteria and other pests; they also induce plant defence mechanisms and stimulate the growth and activity of beneficial microorganisms (Schäfer et al., 2012; Cretoiu et al., 2013; Sharp, 2013), and can also increase the effectiveness of beneficial bacteria against pests (Yu et al., 2008). Chitin fragments are known to have eliciting activities, leading to a variety of defence responses in host plants. Based on these and other properties that help to strengthen host plant defences, interest has been growing in using the technique in agricultural systems to reduce the negative impact of diseases on the yield and quality of crops. When applied as a foliar spray, 0.1% (w/v) chitosan solution increases turmeric plant (*Curcuma longa*

L.) growth parameters, such as shoot height, number of leaves per plant and plant fresh weight, and also has a positive impact on plant yielding and yield quality, including active substances content (Anusuya and Sathiyabama, 2016). Chitosan is especially effective when used in combination with microelements and humic acids, increasing, for example, dry bean (*Phaseolus vulgaris* L.) plant growth parameters and yield quality (Ibrahim and Ramadan, 2015). Recently, positive results with chitin have been obtained in a study conducted by Mahdavi and Rahimi (2013). They tested the effect of 'Chitosan' on the germination and growth of ajowan (*Trachyspermum ammi*). Stimulation of germination and growth was observed with a decrease in the harmful impact of abiotic stress such as high salinity. Significant benefits have been observed for soybean yield, seed germination and plant growth in a study by Zeng et al. (2012), in which soybeans were coated with chitosan. These properties make chitin and its derivatives a useful material for coating seeds used in organic agriculture.

Chitin is also a raw material for the production of innovative liquid organic fertilizers, which are easy to use and effective when applied to plants and soil (Struszczyk et al., 1989). Innovative formulations, like the preparation 'Apol-Humus' allowed for use in organic agriculture in Poland, are a liquid mixture of chitosan and humic acids for foliar and soil application. This preparation stimulates the growth of plants in two ways: through a direct nutritional effect and disease suppression.

6 Products and by-products of plant origin for fertilizers

The 'products and by-products of plant origin' category covers a wide group of leftovers from the food processing industry, for example, oilseed meals, cocoa husks and malt culms. These by-products are sources of slow releasing nitrogen suitable for use in agriculture and horticulture. The levels of nutrients, especially of nitrogen forms, vary according to the source of the raw material and the extent of processing. For example, the average nitrogen content in sunflower meals is about 5.0%, *Brassica* meal 5.7%, neem (*Azadirachta indica*) cake 5.2%, castor cake 4.25% and cotton seed cake 4.6% (Kumar et al., 2014; Mazzoncini et al., 2015). In the case of meals obtained from seeds of some *Brassica* species, there are high levels of glucosinolates, which have an inhibitory effect on nitrification processes (Mazzoncini et al., 2015). The nutritional effect of plant seed meals can be enhanced due to their additional properties, for example, antifungal or zoocidal activity, especially against root-knot nematodes such as *Meloidogyne* (Duong et al., 2014; Sumbul et al., 2015; Yang et al., 2015a).

For fertilization purposes, processed residues of many plant species, especially legumes, are also used. For example, pelletized meal of red clover (Ekofert K) gave the highest marketable yield of celeriac cv. 'Diamant' in comparison with traditional mineral fertilization and non-fertilized control (Kaniszewski et al., 2013).

6.1 Molasses

Another plant product, molasses, is a raw material for the production of some soil improvers and fertilizers. For example, Vinassa is obtained as a by-product during the production of baker's yeast. Vinassa, when applied as a foliar spray, positively affects plant growth and yielding, but its effects are associated with the plant species, variety and also growing conditions such as the type of soil or field localization. For example, large-fruited cranberry (*Vaccinium macrocarpon* L.) cv. 'Stevens' shows good results in terms of plant growth parameters in comparison to 'Pilgrim' and 'Ben Lear' after applications of Vinassa (Derkowska et al., 2015b). In nursery stock production, Vinassa, applied at a rate of 5 L/ha, moderately stimulates the growth of maiden trees, increasing their height and the length of lateral shoots, and has a beneficial effect on the development of the root system and flower buds. Apple trees cv. 'Ariva' produced the best quality fruits as a result of treatment with Vinassa (Rozpara et al., 2014). Also soil microbial community parameters, such as the formation of mycorrhizal structures and the population size of free-living bacteria and fungi communities, positively affected after treatments with Vinassa (Derkowska et al., 2015a; Sas Paszt et al., 2015).

6.2 Natural waste fibres

Natural waste fibres are used as raw materials of plant origin from the textile industry. New products based on these materials are also obtained. These materials are used as a slow release fertilizer when mixed with a source of nitrogen, for example, residues of legume plants. For example, they can be used for the production of agro-fleece for organic vegetable production, which protects crops against weeds, and after crop harvest, the fleece is tilled under and acts as a slow release fertilizer and soil structure improver (Babik et al., 2013). In that case, agro-fleece was enriched with clover leaves as a source of nitrogen, and the obtained yield was similar to that obtained using mineral plant nutrition, but with minimal weed growth in the covered fields.

6.3 Plant-derived hydrolysates

Plants are a source of raw material for the production of plant-derived hydrolysates used in agriculture as bio-stimulants. Hydrolysates of plant origin show not only a nutritional effect but also a hormone-like activity, which results in higher biomass, SPAD index and nutrient content in the tissues of the treated plants (Colla et al., 2014).

'BioFeed Amin', a product based on plant hydrolysates, applied in a dose of 5 L/ha, intensifies the growth of maiden trees, markedly improves their branching ability, stimulates the growth of lateral shoots and improves considerably the quality of the root system. Application of plant-derived hydrolysates has a positive impact on soil microbial communities in comparison with mineral fertilization regimes, resulting in higher number of bacteria in the soil and better mycorrhiza formation in the roots (Sas Paszt et al., 2011; Derkowska, et al., 2015a).

6.4 Seaweeds and seaweed products

Seaweed and seaweed-based products have been used in agriculture for a long time, first as plant material directly applied to the soil, later as solid meals used as fertilizers or soil improvers, with their mode of action based on the minerals released into the soil (Crouch and van Staden, 1993). Seaweeds are a rich source of natural compounds such as amino acids, polysaccharides, lipids, trace elements, vitamins and especially plant growth regulators (auxins, cytokinins, gibberellins, betains, etc.), making seaweed suitable for sustainable, especially organic, agriculture. Seaweed products affect plant growth and biotic and abiotic stress resistance when applied directly onto plants and stimulate the rhizosphere microorganisms when applied to the soil (Crouch et al., 1990; Kuwada et al., 2006; Thorsen et al., 2010; Panda et al., 2012). Seaweed preparations directly affect plant growth and yield quality, both in agricultural and horticultural production (Fan et al., 2011; Hernández-Herrera et al., 2014; Kumar and Sahoo, 2011; Papenfus et al., 2013). In strawberry, for example, the effects of seaweed preparations Kelpak® and Goëmar BM 86® interacted with the plant variety. In the case of 'Elkat', the preparations Kelpak SL and Goëmar BM 86® significantly improved fruit yield, but in the case of 'Salut', there was no effect on yield (Masny et al., 2004). The seaweed preparations Kelpak® and Goëmar BM 86® applied to apple trees, on their own or together in a mixture, stimulated the growth of shoots and leaves and modified fruit coloration, but the effects were also dependent on the cultivar (Basak, 2008). In the case of agricultural crops, applications of the preparation 'Maxicrop' increased shoot and root dry weight of radish in comparison with another seaweed-based product – 'Seasol', and ashes obtained from these products, which were not as effective as Maxicrop (Yusuf et al., 2016). Seaweed extracts can be used to control pests on plants via direct effects or by increasing plant resistance (Hankins and Hockey, 1990; Jayaraj et al., 2008; Jayaraman et al., 2011; Stadnik and Freitas, 2014; Ngala et al., 2016). In the tests conducted in a nursery of the Research Institute of Horticulture in Skierniewice, Poland, the seaweed-based preparation 'BioFeed Quality', applied twice per season at 5 L/ha, intensified the growth of maiden trees, markedly stimulated their branching, caused a significant increase in the length of lateral shoots and had a beneficial effect on the development of the root system of the maiden trees.

In organic agriculture in the EU, only marine seaweed products obtained directly by physical processes including dehydration, freezing, grinding, extraction with water or aqueous acid and/or alkaline solutions, or fermentation can be used. The conditions of extraction have a significant impact on the physical and chemical properties of the final product and its effect on plants under cultivation conditions (Briceño-Domínguez et al., 2014).

6.5 Sawdust and wood chips

Sawdust and wood chips obtained from chemically untreated wood are a soil improver used especially in horticulture. Sawdust, wood chips, wood bark and other similar residues are commonly used as mulch in the cultivation of different horticultural plants, small fruits, vegetables and ornamentals (Haynes and Swift, 1986; Sanderson and Cutcliffe, 1991; Abouziena and Radwan, 2015; Lima et al., 2016). Application of these materials prevents the growth of weeds, but simultaneously 'starves' the cultivated trees of nitrogen, causing their leaves to become yellowish, so that in the following year, the trees grow less vigorously as the nitrogen content in the soil decreases compared to the control (Hoagland et al., 2008).

Another use of sawdust is as a medium or component of media for soilless production of plants, especially in protected (under cover) cultivation, being a promising alternative to the traditional growing media such as rockwool, peat or coconut fibre (Jung et al., 2015; Depardieu et al., 2016). Results of hydroponic production of strawberry fruit on sawdust show that the yield obtained on pure sawdust is lower than that obtained on media based on sawdust mixed with pumice or with coconut fibre (Marinou et al., 2013). Strawberry plants planted in sawdust-based media need additional fertilizer supplementation to satisfy the initial nutrient requirements of bare-root plants (Depardieu et al., 2016).

Sawdust and other wood derivatives are also suitable amendments for manure composting, for the production of composted fertilizers or soil improvers allowed for use in organic agriculture (Cuijpers and Hospers-Brands, 2008; Troy et al., 2012; Khan et al., 2014). Moreover, sawdust and wood chips are used for biochar production (Spokas et al., 2009; Ghani et al., 2013; Laghari et al., 2016).

6.6 Wood ash as fertilizer and soil improver

Wood ash used in organic agriculture must be obtained from wood that has not been treated with chemicals (preservatives, paints, etc.) after felling because this can negatively affect soil quality and plant growth (Jones and Quilliam, 2014). Wood ash contains all the essential plant nutrients, except nitrogen. Wood ash may be directly used to improve crop production and improve soil pH. Wood ash as a natural mineral fertilizer can induce negative priming and promote soil organic matter from grassland soil (Reed et al., 2017). Wood ash reduces fertilization costs and improves crop yields and soil properties, with residual effects observed for up to seven years or more. Wood ash is a good improver during composting of organic matter. Additions of up to 15% do not affect the composting process, but high doses of ash can increase the input of heavy metals present in the ash to limits that are not acceptable for good-quality composts (Fernández-Delgado Juárez et al., 2015).

6.7 Mushroom culture wastes (spent mushroom substrates - SMS)

The composition of this category is limited to products listed in 'Annex II' - 'Products for use in fertilization and soil conditioning'. A typical mushroom substrate is prepared by using straw, poultry litter or horse manure with the addition of gypsum, fermented and inoculated with fungal mycelia. This method is typical for *Agaricus bisporus* cultures. Other cultivable mushrooms, such as *Pleurotus* sp., are cultivated on wood or on cereal straw (Roy et al., 2015). These two kinds of spent substrates differ in nutrient content and physical properties (Paredes et al., 2016). Mushroom cultivation is one of the methods for the utilization of organic residues. However, large amounts of spent substrates are a serious problem for their utilization (Das and Mukherjee, 2007; Koutrotsios et al., 2014). They can be used as raw material in the field, or can be processed to obtain products with better effectiveness.

Spent mushroom substrates or their derivatives promote the growth and yields of plants and increase crop quality. For example, the growth of pepper plants was strongly induced by *Pleurotus* SMS leachate, in comparison with pure or weathered SMS from *Pleurotus* and *Agaricus* cultures (Roy et al., 2015).

6.8 Fermented products

Other traditional products permitted for use in organic agriculture are fermented household wastes or plant matter. In this group, we can also include biogas-digestate-containing plant residues and animal by-products co-digested with materials of plant origin. Agricultural utilization of digestates is one method of managing these by-products (Dahlin et al., 2017).

Biogas digestates may have a beneficial impact on soil properties due to the presence of macro- and microelements and a high organic matter content (Arthurson, 2009; Möller and Müller, 2012). The digestion process and subsequent processing, such as composting, increase the percentage of nitrogen, phosphorus and potassium in the final product (Macias-Corral et al., 2017). Biogas digestates obtained from plants such as maize affect the mineral uptake by crop plants and have an impact on the composition of soil microbial communities (Hupfauf et al., 2016). Biogas digestates used correctly as fertilizer positively impact the growth of cultivated plants, and yields can be similar to those obtained using traditional mineral fertilizers (Barbosa et al., 2014). On the other hand, utilization of the non-separated or liquid phase of biogas plant digestates in high doses may have toxic effects on soil biota and cultivated plants (Stefaniuk et al., 2015).

6.9 Stillage and stillage extract

Materials obtained during ethanol production are suitable as an effective soil amendment in field and horticultural crops. Stillage is a rich source of residual sugars, organic nitrogen forms and other nutrients whose content is associated with the feedstock material used for fermentation (Wilkie et al., 2000; Alotaibi et al., 2014,).

These properties make stillage and its derivatives a good and promising fertilizer, but utilization of raw stillage is restricted by some legal rules due to its potential negative effect on soil properties (Sajbrt et al., 2010; Fuess and Garcia, 2014). Good results with the use of stillage have been obtained in canola and wheat production (Qian et al., 2011; Alotaibi et al., 2014).

6.9.1 Solid phase residues from biogas production

The solid phase of digestate obtained from biogas production also has beneficial properties. Good-quality feedstock allows the obtaining of a digestate suitable for use in agriculture (Al Seadi et al., 2013). Utilization of the separated solid phase of biogas digestate or stillage is one way of alleviating eco-toxic effects of biogas by-products on plants and soil organisms. Application of the solid phase stimulates root growth, whereas liquid phase digestates show inhibiting effects on the growth of roots in some plants (Stefaniuk et al., 2015). Separated, dried solid phase wastes have a higher C:N ratio in comparison with the liquid or non-separated residues. Also their mineral content is associated with the nature of the biogas digestate.

Residues from biogas production can also be used as feedstock for biochar production, but there is a risk of potential toxicity of biochar for soil organisms associated both with the nature of feedstock material production and pyrolysis temperature. For example, Stefaniuk et al. (2016) showed that biochar obtained from non-separated digestate, pyrolysed at 800°C, is much more toxic for soil organisms in comparison with the separated, solid phase of digestate prepared at 400°C.

6.10 Leonardites

Leonardite is naturally oxidized lignite (brown coal), raw organic sediment rich in humic and fulvic acids, mainly obtained as a by-product of near-surface mining activities. It is used as a raw material as a soil conditioner and for the production of pure humic acid preparations, humates (e.g. potassium or iron salts of humic acids) (Kovács et al., 2013). The production of humic acids is conducted with strong alkali or using physical methods (Canieren et al., 2017).

Humic acids stabilize ion exchange in soils and are very useful in contaminated soils. The effects of humic acids on plant growth and nutrient uptake are associated with physical properties of humic acids, especially their molecular weight (Qian et al., 2015; Sun et al., 2016). Leonardites, or humic acids, accelerate microbial activity in soils, including decomposition of organic pollutants (Tejeda-Agredano et al., 2014). Humic substances extracted from leonardites also have bio-stimulant properties due to the presence of polycyclic hydrocarbons, similar in structure to plant growth regulators (Conselvan et al., 2017).

Similar properties as a raw material are exhibited by brown coal (lignite), which is also a good source of humic substances used as soil improvers. Direct application of lignite to the soil is not recommended due to economic considerations, especially the high cost of long-distance transport, in relation to raw lignite costs and the negative impact of raw lignite on soil properties. Possible negative effects on soil microbial communities or nutrient cycling are visible mainly on a short-time scale. For example, lignite added to a sandy or clay soil decreases microbial activity, measured as respiration, and increases the activity of peroxidase and phenol oxidase enzymes (Kim Thi Tran et al., 2015). A much more effective way is to process lignite into concentrated easy-to-use preparations, rich in humic acids. The methods of extraction of humic acids from brown coal are similar to those used for leonardite, but sometimes with additional treatments for better quality and usability of the final product (Doskočil et al., 2014; Ozkan and Ozkan, 2016). Brown coal and peat, mentioned below, are promising sources of humic acids for the organic fertilizer industry (Liu et al., 2014).

6.11 Peat

Peat is a material of moor origin with use limited to horticulture in organic agriculture in the EU, used as a soil conditioner, a growth substrate for plants, or as soil mulch. Peat is also a source of substances that promote plant growth, for example, humic acids, and phytohormone-like organic substances (Klavins and Purmalis, 2013; Boguta et al., 2016; Szajdak, 2016). Peat is used as a raw material for the production of humic acid preparations and fertilizers for use in agriculture and horticulture (Saito and Seckler, 2014; Agafonova et al., 2015). On the EU market, there are some products obtained from peat, for example, a Latvian liquid preparation called Humate Green OK Universal-Pro and similar products by the same producer, prepared on the basis of peat extracts; it can be applied directly to the soil or on plants to increase the nutritional effect.

6.12 Sapropel and similar sediments

Sapropel refers to sediments rich in organic matter formed under exclusion of oxygen in stagnant water bodies. Apart from humic acids, sapropels contain macro- and microelements, simple organic molecules like amino acids, vitamins, antibiotics and other substances that promote the growth of plants (Blečić et al., 2014; Grantina-Ievina et al., 2014; Szajdak, 2016). The amounts of organic substances are associated with the conditions of sapropel formation and the degree of its transformation (Dmitrieva et al., 2015). A characteristic feature of some sapropels is the number of microorganisms that mineralize the organic substances to form sapropel (Tretjakova et al., 2015). These properties make sapropels a good substrate for the production of new organic fertilizers (Agafonova et al., 2015). For example, the sapropel-based organic fertilizer 'Humin Plus' is composed of sapropel, peat and chicken manure, additionally treated for improved effectiveness (Ostrovskij, 2014). Due to the high content of nitrogen in some sapropels, there is a need to determine this content and use sapropels correctly to avoid over-fertilization with nitrogen according to the 91/676/EEB rules (Klepeckas and Januškaitienė, 2017). For organic agriculture use in the EU, only fresh water sediments are allowed. Experiments have shown that sapropel is a good material positively affecting soil quality, plant growth and yield quality parameters. Applications of sapropel increase the chlorophyll content in cereals, such as wheat and barley (Klepeckas and Januškaitienė, 2017). Sapropel-based fertilizer 'Humin Plus' increased the yields of tested crops by 23–25% in wheat, 18–27% in corn, 11–32% in sugar beet, 32% in potato and 45–55% in canola, depending on the location of the experiment (Ostrovskij, 2014).

7 Composts

Composts are produced from separated domestic residues, but there is a possibility to use other substrates like mushroom culture wastes, vegetable matter or other materials alone or mixed with farmyard manure. Composting is also a good way to stabilize nutrient content and dehydration of animal manures. Nowadays, compost is one of the organic fertilizers allowing utilization of organic wastes during the composting process (Yu et al., 2016). Yearly compost applications promote plant growth and reduce the risk of soil and groundwater contamination with nitrates (Toselli et al., 2013). Long-term compost application can increase organic carbon content and improve soil structure, mainly due to the formation of soil macro-aggregates, which are fundamental as habitats for soil microorganisms (Zhang et al., 2014). Effects on yields do depend on the nutrient contents, their availability as well as the crop and its environment. Applications of compost can result in the increase of the yield of some crops, for example, peas (*Pisum sativum* L.), but can also have no impact on the yields of other crops, such as oats (*Avena sativa* L.) (Jannoura et al., 2014).

As a slow nutrient releasing agent, applications of compost as a fertilizer or a component of growing media can increase the yield and quality of harvested fruits and vegetables. The use of compost resulted in a larger size and higher number of 'extra class' strawberry fruits of the cultivars 'Senga Sengana', Kent' and 'Elsanta' in comparison with the control (Frąc et al., 2009). Strawberry plants cv. 'Marmolada' and 'Maya' planted in a compost-peat mixture gave higher yields than those planted in peat; cv. 'Patty' gave higher yields in the peat control, whereas in the case of cv. 'Irma', the effect was correlated with the compost content in the medium – with 25% compost, the yields were better in peat, and with 50% compost in the medium, the yield was lower than that in peat (Altieri et al., 2014).

Compost application has a big impact, mainly positive, on soil microbiota such as soil bacteria or fungi, but compost can also decrease the colonization rate of AMF (Derkowska et al., 2008; Zhen et al., 2014).

Vermicompost is the result of conversion of organic residues by earthworms into humus-like substances. There are many types of raw material used for vermicompost production: precomposted animal faeces, crop residues, food processing by-products, industrial residues and others (Sim and Wu, 2010; Das et al., 2016; Domínguez et al., 2017). Vermicompost can enhance soil fertility physically, chemically and biologically. In physical terms, a vermicompost-treated soil has better aeration, porosity, bulk density, drainage and water retention. Chemical properties such as pH, electrical conductivity and organic matter content are also improved for better crop yield (Lim et al., 2015). Nutrient composition in vermicompost depends on the raw material used as feedstock and the species of earthworms used for the vermicomposting process (Lim et al., 2015; Joshi et al., 2015). Vermicompost is a better nutritional source than traditional composts due to its increased rate of mineralization and degree of humification by the action of earthworms. Nutrients are present in forms readily available to plants. The beneficial properties of vermicomposts are increased by the presence of fungi, bacteria and actinomycetes, producers of plant growth regulators and other substances which can affect plant growth and yields (Joshi et al., 2015).

When applied on their own or with selected strains of beneficial microorganisms, vermicomposts have a positive effect on plant growth and yield (Yang et al., 2015b), and can be a part of sustainable, integrated or certified organic production (Bharti et al., 2016a; Beck et al., 2016).

Vermicomposts are also a raw material for the production of liquid formulations, which can have a positive impact on plant growth, but are easier to use for field applications, especially in perennial, for example, pomological crops (Singh et al., 2010; Grzyb et al., 2012; Rozpara et al., 2014; Mosa et al., 2016). Additions of other natural compounds can improve the survivorship of beneficial bacterial strains in vermicompost-based formulations (Kalra et al., 2010).

8 Untreated minerals and by-products of selected industrial processes

Untreated minerals are not strictly organic fertilizers, but are allowed for application in organic agriculture as a source of phosphorus, potassium, magnesium, calcium and microelements.

This category covers crude potassium salt or kainite, potassium sulphate (K_2SO_4) possibly containing Mg salts, calcium carbonate, magnesium and calcium carbonate, magnesium sulphate (kieserite), calcium chloride solution, calcium sulphate (gypsum), industrial lime from sugar production, industrial lime from vacuum salt production, elemental sulphur and trace elements as inorganic micronutrients listed in EC Regulation 2003/2003 (Ciesielska et al., 2011). Untreated minerals are also used as raw materials for the production of organic fertilizers.

9 Biochar

Biochar is a new promising material for organic agriculture, which can be used as a fertilizer, soil improver or pollutant binder, alone or in a mixture with other amendments. Currently, it is not present on the list of products allowed for use in organic agriculture in the EU. Biochar is a material made from organic matter, pyrolysed at high temperature in the absence of oxygen. Charred materials include materials of plant origin, such as wood residues, crop residues, microalgae biomass, animal residues (bones or meals), animal manure and mixed materials, for example, food industry wastes or sewage sludge (Chan et al., 2007, 2008; Sohi et al., 2010; Bird et al., 2011; Enders et al., 2012; Angst et al., 2014; Hosseini Bai et al., 2015). The chemical and physical characteristics of biochars vary depending on the feedstock material and conditions of the conversion, including temperature, time of processing and the technology used (Lehmann et al., 2011; Cantrell et al., 2012; Baronti et al., 2014; Akhter et al., 2015). Biochars produced in different plants but from the same biomass and under similar pyrolysis conditions can result in various properties of the final product (Spokas et al., 2012a). The final product is a material that can have residual (or relic) structures of the original feedstock material or no residual structures (Spokas, 2010). The physical structure of biochars affects the organic and inorganic composition: the pH can range from 5.6 to 13.0, the C content from 33.0% to 82.7%, N content from 0.1% to 6.0% and the C: N ratio from 19 to 221 (Jha et al., 2010; Spokas, 2010; Spokas et al., 2012b). Biochar can also contain appreciable quantities of P, K, Ca, Mg and micronutrients (Cu, Zn, Fe, Mn), with ashes accounting for 5–60% of the weight, depending on the source of the biomass and pyrolysis conditions (Cheng et al., 2008; Enders et al., 2012).

The main goal of biochar applications in previous years was carbon sequestration in soil deposits (Jha et al., 2010). Now the biochar utilization is also focused on increasing crop yields (Jeffery et al., 2011). The main focus nowadays is also to increase soil fertility (Atkinson et al., 2010) using biochar-induced specific properties of soil and biochar's impact on soil microbiota (Steiner et al., 2008, 2009; Anderson et al., 2011; Parvage et al., 2013; Vanek et al., 2016). Biochar can also be used as a component of growing media for plant production, especially as an alternative to non-renewable materials like peat (Kern et al., 2017). Biochars of animal origin are a suitable source of phosphorus and other macro-elements (Siebers et al., 2014).

10 Conclusion

Utilization of organic or non-processed mineral fertilizers is obligatory in organic agriculture. The wide range of products allows correct soil management and plant fertilization, according to sustainable agricultural practice rules. Organic agriculture is also an area with scope for innovative product creation like organic fertilizers enriched with strains of beneficial microorganisms, for example soil bacteria, filamentous or mycorrhizal fungi. Utilization of organic fertilizers will contribute to the development of organic and sustainable nutrient management strategies and cultivation measures in agriculture.

Organic agriculture also allows better utilization of neglected resources like agricultural or food industry wastes, meat industry residues or biogas station residues. On the market there are a number of products based on processed organic wastes, but utilization rate of these wastes is not adequate to the amount of produced wastes.

Innovations from organic agriculture will spread to other areas of human activity, like bio-fortification of conventionally managed arable fields, restoration of degraded soils and industrial areas or in creation of new green spaces in urbanised areas.

11 Where to look for further information

The present and future research in the area of organic fertilizers will be focused on the utilization of a wide range of natural resources and food industry processing wastes for plant fertilization and soil improvement. The group of industry processing wastes as raw materials for fertilizer production is especially valuable, because its utilization will help to avoid air, water and soil contamination from decayed organic residues. The better utilization of wastes within one process should deliver a raw material for further processes. For example, wastes from biogas production should be used as raw material for production of organic soil improvers. Other wastes, like animal bones or fur feather and horn wastes, can serve as raw materials for the production of slow release of phosphorus or nitrogen fertilizers.

There is a need to develop innovative biofertilizers, new fertilization techniques and nutrient management strategies, which will be applicable in modern, environmentally friendly, sustainable agriculture.

12 References

Abouziena, H. F. and Radwan, S. M. (2015), Allelopathic effects of sawdust, rice straw, bur- clover weed and cogongrass on weed control and development of onion, *International Journal of ChemTech Research*, 7, pp. 337–45.

Agafonova, L., Alsina, I., Sokolov, G., Kovrik, S., Bambalov, N., Apse, J. and Rak, M. (2015), New kinds of sapropel and peat based fertilizers, in: *Environment. Technology. Resources*. Rezekne, Latvia, Volume 2, pp. 20–6.

Ahmad, F., Ahmad, I. and Khan, M. S. (2008), Screening of free-living rhizospheric bacteria for their multiple plant growth promoting activities, *Microbiological Research*, 163(2), pp. 173–81. DOI:10.1016/j.micres.2006.04.001.

Akhter, A., Hage-Ahmed, K., Soja, G. and Steinkellner, S. (2015), Compost and biochar alter mycorrhization, tomato root exudation, and development of *Fusarium oxysporum* f. sp. *lycopersici*, *Frontiers in Plant Science*, 6, pp. 529. DOI:10.3389/fpls.2015.00529.

Al Seadi, T., Drosg, B., Fuchs, W., Rutz, D. and Janssen, R. (2013), Biogas digestate quality and utilization, in: Murphy, J. and Baxter, D. (Eds), *The Biogas Handbook: Science, Production and Applications*. Sawston/Cambridge, UK: Woodhead Publishing, pp. 267–301.

Alotaibi, K. D., Schoenau, J. J. and Hao, X. (2014), Fertilizer potential of thin stillage from wheat-based ethanol production, *BioEnergy Research*, 7(4), pp. 1421–9. DOI:10.1007/s12155-014-9473-1.

Altieri, R., Esposito, A., Baruzzi, G. and Nair, T. (2014), Corroboration for the successful application of humified olive mill waste compost in soilless cultivation of strawberry, *International Biodeterioration & Biodegradation*, 88, pp. 118–24. DOI:10.1016/j.ibiod.2013.12.006.

Anderson, C. R., Condron, L. M., Clough, T. J., Fiers, M., Stewart, A., Hill, R. A. and Sherlock, R. R. (2011), Biochar induced soil microbial community change: Implications for biogeochemical cycling of carbon, nitrogen and phosphorus, *Pedobiologia*, 54(5-6), pp. 309–20. DOI:10.1016/j.pedobi.2011.07.005.

Angst, T. E., Six, J., Reay, D. S. and Sohi, S. P. (2014), Impact of pine chip biochar on trace greenhouse gas emissions and soil nutrient dynamics in an annual ryegrass system in California, *Environmental Benefits and Risks of Biochar Application to Soil*, 191, pp. 17-26. DOI:10.1016/j.agee.2014.03.009.

Anusuya, S. and Sathiyabama, M. (2016), Effect of chitosan on growth, yield and curcumin content in turmeric under field condition, *Biocatalysis and Agricultural Biotechnology*, 6, pp. 102-6. DOI:10.1016/j.bcab.2016.03.002.

Arthurson, V. (2009), Closing the global energy and nutrient cycles through application of biogas residue to agricultural land-potential benefits and drawback, *Energies*, 2(2), pp. 226-42.

Arun, K. S. (2007), *Bio-fertilizers for Sustainable Agriculture. Mechanism of P-Solubilization*. Sixth Edition, Jodhpur, India: Agribios Publishers, pp. 196-7.

Atkinson, C. J., Fitzgerald, J. D. and Hipps, N. A. (2010), Potential mechanisms for achieving agricultural benefits from biochar application to temperate soils: A review, *Plant and Soil*, 337(1), pp. 1-18. DOI:10.1007/s11104-010-0464-5.

Aznar, A., Chen, N. W. G., Rigault, M., Riache, N., Joseph, D., Desmaële, D., Mouille, G., Boutet, S., Soubigou-Taconnat, L., Renou, J. P., Thomine, S., Expert, D. and Dellagi, A. (2014), Scavenging iron: A novel mechanism of plant immunity activation by microbial siderophores, *Plant Physiology*, 164(4), pp. 2167-83. DOI:10.1104/pp.113.233585.

Babik, J., Babik, I. and Kaniszewski, S. (2013), New biodegradable agro-fleece for soil mulching in organic vegetable production, in: De Neve, S. (Ed.), *NUTRIHORT: Nutrient Management, Innovative Techniques and Nutrient Legislation in Intensive Horticulture for an Improved Water Quality*. Ghent, Netherlands, pp. 343-9.

Barbosa, D. B. P., Nabel, M. and Jablonowski, N. D. (2014), Biogas-digestate as nutrient source for biomass production of *Sida hermaphrodita*, *Zea mays* L. and *Medicago sativa* L., *Energy Procedia*, 59, pp. 120-6. European Geosciences Union General Assembly 2014, EGU Division Energy, Resources & the Environment (ERE). DOI:10.1016/j.egypro.2014.10.357.

Baronti, S., Vaccari, F. P., Miglietta, F., Calzolari, C., Lugato, E., Orlandini, S., Pini, R., Zulian, C. and Genesio, L. (2014), Impact of biochar application on plant water relations in *Vitis vinifera* (L.), *European Journal of Agronomy*, 53, pp. 38-44. DOI:10.1016/j.eja.2013.11.003.

Basak, A. (2008), Effect of preharvest treatment with seaweed products, Kelpak® and Goëmar BM 86®, on fruit quality in apple, *International Journal of Fruit Science*, 8(1-2), pp. 1-14. DOI:10.1080/15538360802365251.

Beck, J. E., Schroeder-Moreno, M. S., Fernandez, G. E., Grossman, J. M. and Creamer, N. G. (2016), Effects of cover crops, compost, and vermicompost on strawberry yields and nitrogen availability in North Carolina, *HortTechnology*, 26(5), pp. 604-13. DOI:10.21273/HORTTECH03447-16.

Bedini, S., Turrini, A., Rigo, C., Argese, E. and Giovannetti, M. (2010), Molecular characterization and glomalin production of arbuscular mycorrhizal fungi colonizing a heavy metal polluted ash disposal island, downtown Venice, *Soil Biology and Biochemistry*, 42(5), pp. 758-65. DOI:10.1016/j.soilbio.2010.01.010.

Bharti, N., Barnawal, D., Shukla, S., Tewari, S. K., Katiyar, R. S. and Kalra, A. (2016a), Integrated application of *Exiguobacterium oxidotolerans*, *Glomus fasciculatum*, and vermicompost improves growth, yield and quality of *Mentha arvensis* in salt-stressed soils, *Industrial Crops and Products*, 83, pp. 717–28. DOI:10.1016/j. indcrop.2015.12.021.

Bharti, N., Barnawal, D., Wasnik, K., Tewari, S. K. and Kalra, A. (2016b), Co-inoculation of *Dietzia natronolimnaea* and *Glomus intraradices* with vermicompost positively influences *Ocimum basilicum* growth and resident microbial community structure in salt affected low fertility soils, *Applied Soil Ecology*, 100, pp. 211–25. DOI:10.1016/j. apsoil.2016.01.003.

Bird, M. I., Wurster, C. M., de Paula Silva, P. H., Bass, A. M. and de Nys, R. (2011), Algal biochar – production and properties, *Bioresource Technology*, 102(2), pp. 1886–91. DOI:10.1016/j.biortech.2010.07.106.

Blečić, A., Railić, B., Dubljević, R., Mitrović, D. and Spalevic, V. (2014), Application of sapropel in agricultural production, *Agriculture and Forestry*, 60(2), pp. 243–50.

Boguta, P., D'Orazio, V., Sokołowska, Z. and Senesi, N. (2016), Effects of selected chemical and physicochemical properties of humic acids from peat soils on their interaction mechanisms with copper ions at various pHs, *Journal of Geochemical Exploration*, 168, pp. 119–26. DOI:10.1016/j.gexplo.2016.06.004.

Bona, E., Lingua, G., Manassero, P., Cantamessa, S., Marsano, F., Todeschini, V., Copetta, A., D'Agostino, G., Massa, N., Avidano, L., Gamalero, E. and Berta, G. (2015), AM fungi and PGP pseudomonads increase flowering, fruit production, and vitamin content in strawberry grown at low nitrogen and phosphorus levels, *Mycorrhiza*, 25(3), pp. 181–93. DOI:10.1007/s00572-014-0599-y.

Bona, E., Cantamessa, S., Massa, N., Manassero, P., Marsano, F., Copetta, A., Lingua, G., D'Agostino, G., Gamalero, E. and Berta, G. (2017), Arbuscular mycorrhizal fungi and plant growth-promoting pseudomonads improve yield, quality and nutritional value of tomato: A field study, *Mycorrhiza*, 27(1), pp. 1–11. DOI:10.1007/ s00572-016-0727-y.

Born, Y., Remus-Emsermann, M. N., Bieri, M., Kamber, T., Piel, J. and Pelludat, C. (2016), Fe^{2+} chelator proferrorosamine A: A gene cluster of Erwinia rhapontici P45 involved in its synthesis and its impact on growth of Erwinia amylovora CFBP1430, *Microbiology*, 162(2), pp. 236–45.

Boursier, H., Béline, F. and Paul, E. (2005), Piggery wastewater characterisation for biological nitrogen removal process design, *Bioresource Technology*, 96(3), pp. 351–8. DOI:10.1016/j.biortech.2004.03.007.

Briceño-Domínguez, D., Hernández-Carmona, G., Moyo, M., Stirk, W. and van Staden, J. (2014), Plant growth promoting activity of seaweed liquid extracts produced from Macrocystis pyrifera under different pH and temperature conditions, *Journal of Applied Phycology*, 26(5), pp. 2203–10. DOI:10.1007/s10811-014-0237-2.

Canieren, O., Karaguzel, C. and Aydin, A. (2017), Effect of physical pre-enrichment on humic substance recovery from leonardite, *Physicochemical Problems of Mineral Processing*, 53(1), pp. 502–14.

Cantrell, K. B., Hunt, P. G., Uchimiya, M., Novak, J. M. and Ro, K. S. (2012), Impact of pyrolysis temperature and manure source on physicochemical characteristics of biochar, *Bioresource Technology*, 107, pp. 419–28. DOI:10.1016/j.biortech.2011.11.084.

Cassán, F., Vanderleyden, J. and Spaepen, S. (2014), Physiological and agronomical aspects of phytohormone production by model Plant-Growth-Promoting Rhizobacteria (PGPR) belonging to the genus Azospirillum, *Journal of Plant Growth Regulation*, 33(2), pp. 440–59. DOI:10.1007/s00344-013-9362-4.

Cavagnaro, T. R., Bender, S. F., Asghari, H. R. and Heijden, M. G. A. van der (2015), The role of arbuscular mycorrhizas in reducing soil nutrient loss, *Trends in Plant Science*, 20(5), pp. 283–90. DOI:10.1016/j.tplants.2015.03.004.

Chan, K. Y., Van Zwieten, L., Meszaros, I., Downie, A. and Joseph, S. (2007), Agronomic values of greenwaste biochar as a soil amendment, *Soil Research*, 45(8), pp. 629–34.

Chan, K. Y., Van Zwieten, L., Meszaros, I., Downie, A. and Joseph, S. (2008), Using poultry litter biochars as soil amendments, *Soil Research*, 46(5), pp. 437–44.

Chantal, K., Shao, X., Jing, B., Yuan, Y., Hou, M. and Liao, L. (2013), Effects of effective microorganisms (EM) and bio-organic fertilizers on growth parameters and yield quality of flue-cured tobacco (Nicotiana tabacum), *Journal of Food, Agriculture and Environment*, 11(2), pp. 1212–15.

Chen, Y. P., Rekha, P. D., Arun, A. B., Shen, F. T., Lai, W.-A. and Young, C. C. (2006), Phosphate solubilizing bacteria from subtropical soil and their tricalcium phosphate solubilizing abilities, *Applied Soil Ecology*, 34(1), pp. 33–41. DOI:10.1016/j.apsoil.2005.12.002.

Cheng, C.-H., Lehmann, J., Thies, J. E. and Burton, S. D. (2008), Stability of black carbon in soils across a climatic gradient, *Journal of Geophysical Research: Biogeosciences*, 113(G2), pp. G02027. DOI:10.1029/2007JG000642.

Ciesielska, J., Malusa, E. and Sas-Paszt, L. (2011), Środki ochrony roślin stosowane w rolnictwie ekologicznym, *Komentarz Do Załącznika II Rozporządzenia Komisji (WE) Nr*, 889(2008), pp. 17–19.

Chetverikov, S. P. and Loginov, O. N. (2009), New metabolites of Azotobacter vinelandii exhibiting antifungal activity, *Microbiology*, 78(4), pp. 428–32. DOI:10.1134/S0026261709040055.

Colla, G., Rouphael, Y., Canaguier, R., Svecova, E. and Cardarelli, M. (2014), Biostimulant action of a plant-derived protein hydrolysate produced through enzymatic hydrolysis, *Frontiers in Plant Science*, 5, pp. 448. DOI:10.3389/fpls.2014.00448.

Colla, G., Nardi, S., Cardarelli, M., Ertani, A., Lucini, L., Canaguier, R. and Rouphael, Y. (2015), Protein hydrolysates as biostimulants in horticulture, *Biostimulants in Horticulture*, 196, pp. 28–38. DOI:10.1016/j.scienta.2015.08.037.

Conselvan, G. B., Pizzeghello, D., Francioso, O., Di Foggia, M., Nardi, S. and Carletti, P. (2017), Biostimulant activity of humic substances extracted from leonardites, *Plant Soil*, 420(1–2), pp. 119–34. DOI:10.1007/s11104-017-3373-z.

Corte, L., Dell'Abate, M. T., Magini, A., Migliore, M., Felici, B., Roscini, L., Sardella, R., Tancini, B., Emiliani, C., Cardinali, G. and Benedetti, A. (2014), Assessment of safety and efficiency of nitrogen organic fertilizers from animal-based protein hydrolysates—a laboratory multidisciplinary approach, *Journal of the Science of Food and Agriculture*, 94(2), pp. 235–45. DOI:10.1002/jsfa.6239.

Cretoiu, M. S., Korthals, G. W., Visser, J. H. M. and van Elsas, J. D. (2013), Chitin amendment increases soil suppressiveness toward plant pathogens and modulates the actinobacterial and oxalobacteraceal communities in an experimental agricultural field, *Applied and Environmental Microbiology*, 79(17), pp. 5291–301. DOI:10.1128/AEM.01361-13.

Crouch, I. J. and van Staden, J. (1993), Evidence for the presence of plant growth regulators in commercial seaweed products, *Plant Growth Regulation*, 13(1), pp. 21–9. DOI:10.1007/BF00207588.

Crouch, I. J., Beckett, R. P. and Staden, J. (1990), Effect of seaweed concentrate on the growth and mineral nutrition of nutrient-stressed lettuce, *Journal of Applied Phycology*, 2(3), pp. 269–72. DOI:10.1007/BF02179784.

Cuijpers, W. J. M. and Hospers-Brands, A. (2008), *Hulpmeststoffen: beschikbaarheid en opname van stikstof in de biologische teelt van zomertarwe*. Driebergen, the Netherlands: Louis Bolk Instituut, 33pp.

Dahlin, J., Nelles, M. and Herbes, C. (2017), Biogas digestate management: Evaluating the attitudes and perceptions of German gardeners towards digestate-based soil amendments, *Resources, Conservation and Recycling*, 118, pp. 27–38. DOI:10.1016/j. resconrec.2016.11.020.

Das, N. and Mukherjee, M. (2007), Cultivation of Pleurotus ostreatus on weed plants, *Bioresource Technology*, 98(14), pp. 2723–6. DOI:10.1016/j.biortech.2006.09.061.

Das, V., Satyanarayan, S. and Satyanarayan, S. (2016), Value added product recovery from sludge generated during gum arabic refining process by vermicomposting, *Environmental Monitoring and Assessment*, 188(9), pp. 523. DOI:10.1007/s10661-016-5516-8.

Davison, J., Moora, M., Öpik, M., Adholeya, A., Ainsaar, L., Bâ, A., Burla, S., Diedhiou, A. G., Hiiesalu, I., Jairus, T., Johnson, N. C., Kane, A., Koorem, K., Kochar, M., Ndiaye, C., Pärtel, M., Reier, Ü., Saks, Ü., Singh, R., Vasar, M. and Zobel, M. (2015), Global assessment of arbuscular mycorrhizal fungus diversity reveals very low endemism, *Science*, 349(6251), pp. 970. DOI:10.1126/science.aab1161.

Depardieu, C., Prémont, V., Boily, C. and Caron, J. (2016), Sawdust and bark-based substrates for soilless strawberry production: Irrigation and electrical conductivity management, *PLoS ONE*, 11(4), pp. e0154104 (1–20). DOI:10.1371/journal. pone.0154104.

Derkowska, E., Sas-Paszt, L., Sumorok, B., Szwonek, E. and Gluszek, S. (2008), The influence of mycorrhization and organic mulches on mycorrhizal frequency in apple and strawberry roots, *Journal of Fruit and Ornamental Plant Research*, 16, pp. 227–42.

Derkowska, E., Sas Paszt, L. S., Harbuzov, A. and Sumorok, B. (2015a), Root growth, mycorrhizal frequency and soil microorganisms in strawberry as affected by biopreparations, *Advances in Microbiology*, 5(1), pp. 65–73.

Derkowska, E., Sas Paszt, L. and Szwonek, E. (2015b), Influence of biological products on the growth and development of large-fruited cranberry under greenhouse conditions, *Folia Horticulturae*, 27(1), pp. 71–7. DOI:10.1515/fhort-2015-0016.

Dhar, D. W., Prasanna, R., Pabbi, S. and Vishwakarma, R. (2015), Significance of cyanobacteria as inoculants in agriculture, in: Das, D. (Ed.), *Algal Biorefinery: An Integrated Approach*. Cham, Switzerland: Springer International Publishing, pp. 339–74.

Di Stasio, E., Maggio, A., Ventorino, V., Pepe, O., Raimondi, G. and De Pascale, S. (2017), Free-living (N2)-fixing bacteria as potential enhancers of tomato growth under salt stress. *Acta Horticulturae*, 1164, pp. 151–6. DOI:10.17660/ActaHortic.2017.1164.19

Dmitrieva, E., Efimova, E., Siundiukova, K. and Perelomov, L. (2015), Surface properties of humic acids from peat and sapropel of increasing transformation, *Environmental Chemistry Letters*, 13(2), pp. 197–202. DOI:10.1007/s10311-015-0497-3.

Domínguez, J., Sanchez-Hernandez, J. C. and Lores, M. (2017), Vermicomposting of winemaking by-products, in: Galanakis, C. M. (Ed.), *Handbook of Grape Processing By-Products*. London, UK: Academic Press, Elsevier, pp. 55–78.

Doskočil, L., Grasset, L., Válková, D. and Pekař, M. (2014), Hydrogen peroxide oxidation of humic acids and lignite, *Fuel*, 134, pp. 406–13. DOI:10.1016/j.fuel.2014.06.011.

Duong, D. H., Ngo, X. Q., Do, D. G., Le, T. A. H., Nguyen, V. T. and Nic, S. (2014), Effective control of neem (*Azadirachta indica* A. Juss) cake to plant parasitic nematodes and fungi in black pepper diseases in vitro, *Journal of Vietnamese Environment*, 6 (3), pp. 233–8.

Ebrahiminezhad, A., Manafi, Z., Berenjian, A., Kianpour, S. and Ghasemi, Y. (2017), Iron-reducing bacteria and iron nanostructures, *Journal of Advanced Medical Sciences and Applied Technologies*, 3(1), pp. 9–16.

Egamberdiyeva, D. (2007), The effect of plant growth promoting bacteria on growth and nutrient uptake of maize in two different soils, *Applied Soil Ecology*, 36(2–3), pp. 184–9. DOI:10.1016/j.apsoil.2007.02.005.

Emerson, J. K. and Roark, A. M. (2007), Composition of guano produced by frugivorous, sanguivorous, and insectivorous bats, *Acta Chiropterologica*, 9(1), pp. 261–7. DOI:10.3161/1733-5329(2007)9[261:COGPBF]2.0.CO;2.

Enders, A., Hanley, K., Whitman, T., Joseph, S. and Lehmann, J. (2012), Characterization of biochars to evaluate recalcitrance and agronomic performance, *Bioresource Technology*, 114, pp. 644–53. DOI:10.1016/j.biortech.2012.03.022.

Fan, D., Hodges, D. M., Zhang, J., Kirby, C. W., Ji, X., Locke, S. J., Critchley, A. T. and Prithiviraj, B. (2011), Commercial extract of the brown seaweed *Ascophyllum nodosum* enhances phenolic antioxidant content of spinach (*Spinacia oleracea* L.) which protects *Caenorhabditis elegans* against oxidative and thermal stress, *Food Chemistry*, 124(1), pp. 195–202. DOI:10.1016/j.foodchem.2010.06.008.

Fernández-Delgado Juárez, M., Gómez-Brandón, M. and Insam, H. (2015), Merging two waste streams, wood ash and biowaste, results in improved composting process and end products, *Science of the Total Environment*, 511, pp. 91–100. DOI:10.1016/j.scitotenv.2014.12.037.

Fox, A. R., Soto, G., Valverde, C., Russo, D., Lagares, A., Zorreguieta, Á., Alleva, K., Pascuan, C., Frare, R., Mercado-Blanco, J., Dixon, R. and Ayub, N. D. (2016), Major cereal crops benefit from biological nitrogen fixation when inoculated with the nitrogen-fixing bacterium Pseudomonas protegens Pf- 5 X940, *Environmental Microbiology*, 18(10), pp. 3522–34. DOI:10.1111/1462-2920.13376.

Frąc, M., Michalski, P. and Sas Paszt, L. (2009), The effect of mulch and mycorrhiza on fruit yield and size of three strawberry cultivars, *Journal of Fruit and Ornamental Plant Research*, 17(2), pp. 85–93.

Fuess, L. T. and Garcia, M. L. (2014), Implications of stillage land disposal: A critical review on the impacts of fertigation, *Journal of Environmental Management*, 145, pp. 210–29. DOI:10.1016/j.jenvman.2014.07.003.

Ghani, W. A. W. A. K., Mohd, A., da Silva, G., Bachmann, R. T., Taufiq-Yap, Y. H., Rashid, U. and Al-Muhtaseb, A. H. (2013), Biochar production from waste rubber-wood-sawdust and its potential use in C sequestration: Chemical and physical characterization, *Industrial Crops and Products*, 44, pp. 18–24. DOI:10.1016/j.indcrop.2012.10.017.

Ghasemi, S., Khoshgoftarmanesh, A., Hadadzadeh, H. and Afyuni, M. (2013), Synthesis, characterization, and theoretical and experimental investigations of zinc(II)–amino acid complexes as ecofriendly plant growth promoters and highly bioavailable sources of zinc, *Journal of Plant Growth Regulation*, 32(2), pp. 315–23. DOI:10.1007/s00344-012-9300-x.

Glinicki, R., Sas-Paszt, L. and Jadczuk-Tobjasz, E. (2011), The effect of microbial inoculation with EM-Farming inoculum on the vegetative growth of three strawberry cultivars, *Annals of Warsaw University of Life Sciences- SGGW. Horticulture and Landscape Architecture*, 32, pp. 3–14.

Grantina-Ievina, L., Karlsons, A., Andersone-Ozola, U. and Ievinsh, G. (2014), Effect of freshwater sapropel on plants in respect to its growth-affecting activity and cultivable microorganism content, *Zemdirbyste-Agriculture*, 101(4), pp. 355–66.

Grzyb, Z. S., Piotrowski, W., Bielicki, P., Sas Paszt, L. and Malusà, E. (2012), Effect of different fertilizers and amendments on the growth of apple and sour cherry rootstocks in an organic nursery, *Journal of Fruit and Ornamental Plant Research*, 20(1), pp. 43–53. DOI:10.2478/v10290-012-0004-x.

Grzyb, Z. S., Sas Paszt, L., Piotrowski, W. and Malusa, E. (2015), The influence of mycorrhizal fungi on the growth of apple and sour cherry maidens fertilized with different bioproducts in the organic nursery, *Journal of Life Sciences*, 9, pp. 221–8.

Gusain, Y. S., Kamal, R., Mehta, C. M., Singh, U. S. and Sharma, A. K. (2015), Phosphate solubilizing and indole- 3-acetic acid producing bacteria from the soil of Garhwal Himalaya aimed to improve the growth of rice, *Journal of Environmental Biology*, 36(1), pp. 301.

Hankins, S. D. and Hockey, H. P. (1990), The effect of a liquid seaweed extract from *Ascophyllum nodosum* (Fucales, Phaeophyta) on the two-spotted red spider mite *Tetranychus urticae*, *Hydrobiologia*, 204–205(1), pp. 555–9. DOI:10.1007/BF00040286.

Haynes, R. J. and Swift, R. S. (1986), Effect of soil amendments and sawdust mulching on growth, yield and leaf nutrient content of highbush blueberry plants, *Scientia Horticulturae*, 29(3), pp. 229–38. DOI:10.1016/0304-4238(86)90066-X.

Hernández-Herrera, R. M., Santacruz-Ruvalcaba, F., Ruiz-López, M. A., Norrie, J. and Hernández-Carmona, G. (2014), Effect of liquid seaweed extracts on growth of tomato seedlings (*Solanum lycopersicum* L.), *Journal of Applied Phycology*, 26(1), pp. 619–28. DOI:10.1007/s10811-013-0078-4.

Hoagland, L., Carpenter-Boggs, L., Granatstein, D., Mazzola, M., Smith, J., Peryea, F. and Reganold, J. P. (2008), Orchard floor management effects on nitrogen fertility and soil biological activity in a newly established organic apple orchard, *Biology and Fertility of Soils*, 45(1), pp. 11. DOI:10.1007/s00374-008-0304-4.

Hosseini Bai, S., Xu, C.-Y., Xu, Z., Blumfield, T., Zhao, H., Wallace, H., Reverchon, F. and Van Zwieten, L. (2015), Soil and foliar nutrient and nitrogen isotope composition (δ15N) at 5 years after poultry litter and green waste biochar amendment in a macadamia orchard, *Environmental Science and Pollution Research*, 22(5), pp. 3803–9. DOI:10.1007/s11356-014-3649-2.

Hu, C. and Qi, Y. (2013), Long-term effective microorganisms application promote growth and increase yields and nutrition of wheat in China, *European Journal of Agronomy*, 46, pp. 63–7. DOI:10.1016/j.eja.2012.12.003.

Hupfauf, S., Bachmann, S., Fernández-Delgado Juárez, M., Insam, H. and Eichler-Löbermann, B. (2016), Biogas digestates affect crop P uptake and soil microbial community composition, 542, pp. 1144–54. Special Issue on Sustainable Phosphorus Taking Stock: Phosphorus Supply from Natural and Anthropogenic Pools in the 21st Century. DOI:10.1016/j.scitotenv.2015.09.025.

Ibrahim, E. A. and Ramadan, W. A. (2015), Effect of zinc foliar spray alone and combined with humic acid or/and chitosan on growth, nutrient elements content and yield of dry bean (*Phaseolus vulgaris* L.) plants sown at different dates, *Scientia Horticulturae*, 184, pp. 101–5. DOI:10.1016/j.scienta.2014.11.010.

Jakienė, E. (2013), The effect of the microelement fertilizers and biological preparation Terra Sorb Foliar on spring rape crop, *Žemės Ūkio Mokslai*, 20(2), pp. 75-83.

Jannoura, R., Joergensen, R. G. and Bruns, C. (2014), Organic fertilizer effects on growth, crop yield, and soil microbial biomass indices in sole and intercropped peas and oats under organic farming conditions, *European Journal of Agronomy*, 52(Part B), pp. 259-70. DOI:10.1016/j.eja.2013.09.001.

Jayaraj, J., Wan, A., Rahman, M. and Punja, Z. K. (2008), Seaweed extract reduces foliar fungal diseases on carrot, *Crop Protection*, 27 (10), pp. 1360-6. DOI:10.1016/j.cropro.2008.05.005.

Jayaraman, J., Norrie, J. and Punja, Z. (2011), Commercial extract from the brown seaweed *Ascophyllum nodosum* reduces fungal diseases in greenhouse cucumber, *Journal of Applied Phycology*, 23(3), pp. 353-61. DOI:10.1007/s10811-010-9547-1.

Jeffery, S., Verheijen, F. G. A., van der Velde, M. and Bastos, A. C. (2011), A quantitative review of the effects of biochar application to soils on crop productivity using meta-analysis, *Agriculture, Ecosystems & Environment*, 144(1), pp. 175-87. DOI:10.1016/j.agee.2011.08.015.

Jha, P., Biswas, A. K., Lakaria, B. L. and Rao, A. S. (2010), Biochar in agriculture-prospects and related implications, *Current Science*, 99(9), pp. 1218-25.

Ji, R., Dong, G., Shi, W. and Min, J. (2017), Effects of liquid organic fertilizers on plant growth and rhizosphere soil characteristics of Chrysanthemum, *Sustainability*, 9(5), pp. 841 (1-16). DOI:10.3390/su9050841.

Jones, D. L. and Quilliam, R. S. (2014), Metal contaminated biochar and wood ash negatively affect plant growth and soil quality after land application, *Journal of Hazardous Materials*, 276, pp. 362-70. DOI:10.1016/j.jhazmat.2014.05.053.

Jones, S. B., Robbins, C. W. and Hansen, C. L. (1993), Sodic soil reclamation using cottage cheese (acid) whey, *Arid Soil Research and Rehabilitation*, 7(1), pp. 51-61. DOI:10.1080/15324989309381334.

Joshi, R., Singh, J. and Vig, A. (2015), Vermicompost as an effective organic fertilizer and biocontrol agent: Effect on growth, yield and quality of plants, *Reviews in Environmental Science and Bio/Technology*, 14(1), pp. 137-59. DOI:10.1007/s11157-014-9347-1.

Jung, J. Y., Kim, J. S., Ha, S. Y., Choi, J. H. and Yang, J.-K. (2015), Suitability of thermal treated sawdust as replacements for peat moss in horticultural media, *Journal of Agriculture and Life Science*, 49(4), pp. 105-15.

Kalra, A., Chandra, M., Awasthi, A., Singh, A. K. and Khanuja, S. P. S. (2010), Natural compounds enhancing growth and survival of rhizobial inoculants in vermicompost-based formulations, *Biology and Fertility of Soils*, 46(5), pp. 521-4. DOI:10.1007/s00374-010-0443-2.

Kaniszewski, S., Babik, I. and Babik, J. (2013), Pelletized legume plants as fertilizer for vegetables in organic farming, in: De Neve, S. (Ed.), *NUTRIHORT: Nutrient Management, Innovative Techniques and Nutrient Legislation in Intensive Horticulture for an Improved Water Quality*. Ghent, Netherlans, pp. 330-42.

Kern, J., Tammeorg, P., Shanskiy, M., Sakrabani, R., Knicker, H., Kammann, C., Tuhkanen, E.-M., Smidt, G., Prasad, M., Tiilikkala, K., Sohi, S., Gascó, G., Steiner, C. and Glaser, B. (2017), Synergistic use of peat and charred material in growing media - an option to reduce the pressure on peatlands? *Journal of Environmental Engineering and Landscape Management*, 25(2), pp. 160-74. DOI:10.3846/16486897.2017.128466 5.

Khan, W., Rayirath, U., Subramanian, S., Jithesh, M., Rayorath, P., Hodges, D., Critchley, A. T., Craigie, J. S., Norrie, J. and Prithiviraj, B. (2009), Seaweed extracts as biostimulants of plant growth and development, *Journal of Plant Growth Regulation*, 28 (4), pp. 386-99.

Khan, N., Clark, I., Sánchez-Monedero, M. A., Shea, S., Meier, S. and Bolan, N. (2014), Maturity indices in co- composting of chicken manure and sawdust with biochar, *Bioresource Technology*, 168, pp. 245-51. DOI:10.1016/j.biortech.2014.02.123.

Kiers, E. T., Rousseau, R. A., West, S. A. and Denison, R. F. (2003), Host sanctions and the legume-rhizobium mutualism, *Nature*, 425(6953), pp. 78-81. DOI:10.1038/nature01931.

Kim Thi Tran, C., Rose, M. T., Cavagnaro, T. R. and Patti, A. F. (2015), Lignite amendment has limited impacts on soil microbial communities and mineral nitrogen availability, *Applied Soil Ecology*, 95, pp. 140-50. DOI:10.1016/j.apsoil.2015.06.020.

Klavins, M. and Purmalis, O. (2013), Properties and structure of raised bog peat humic acids, *Journal of Molecular Structure*, 1050, pp. 103-13. DOI:10.1016/j.molstruc.2013.07.021.

Kleiber, T., Starzyk, J., Górski, R., Sobieralski, K., Siwulski, M., Rempulska, A. and Sobiak, A. (2014), The studies on applying of Effective Microorganisms (EM) and CRF on nutrient contents in leaves and yielding of tomato, *Acta Scientiarum Polonorum - Hortorum Cultus*, 13 (1), pp. 79-90.

Klepeckas, M. and Januškaitienė, I. (2017), Changes in Triticum aestivum and Hordeum vulgare chlorophyll content and fluorescence parameters under impact of various sapropel concentrations, *Biologija*, 62 (4), pp. 216-26.

Köpke, U., Krämer, J. and Leifert, C. (2007), Pre-harvest strategies to ensure the microbiological safety of fruit and vegetables from manure-based production systems, in: Cooper, J., Niggli, U. and Leifert, C. (Eds), *Handbook of Organic Food Safety and Quality*. Cambridge, UK: Woodhead Publishing Ltd., pp. 413-29.

Koutrotsios, G., Mountzouris, K. C., Chatzipavlidis, I. and Zervakis, G. I. (2014), Bioconversion of lignocellulosic residues by Agrocybe cylindracea and Pleurotus ostreatus mushroom fungi – assessment of their effect on the final product and spent substrate properties, *Food Chemistry*, 161, pp. 127-35. DOI:10.1016/j.foodchem.2014.03.121.

Kovács, K., Czech, V., Fodor, F., Solti, A., Lucena, J. J., Santos-Rosell, S. and Hernández-Apaolaza, L. (2013), Characterization of Fe-Leonardite complexes as novel natural iron fertilizers, *Journal of Agricultural and Food Chemistry*, 61(50), pp. 12200-10. DOI:10.1021/jf404455y.

Kumar, G. and Sahoo, D. (2011), Effect of seaweed liquid extract on growth and yield of Triticum aestivum var. Pusa Gold, *Journal of Applied Phycology*, 23(2), pp. 251-5. DOI:10.1007/s10811-011-9660-9.

Kumar, V., Kumar, S., Jha, S. K. and Jijeesh, C. M. (2014), Influence of de-oiled seed cakes on seedling performance of East Indian Rosewood (Dalbergia latifoila Roxb.), *Soil Environment*, 33(2), pp. 169-74.

Kumar, V., Menon, S., Agarwal, H. and Gopalakrishnan, D. (2017), Characterization and optimization of bacterium isolated from soil samples for the production of siderophores, *Resource-Efficient Technologies*, 3(4), 434-9. DOI:10.1016/j.reffit.2017.04.004.

Kuwada, K., Kuramoto, M., Utamura, M., Matsushita, I. and Ishii, T. (2006), Isolation and structural elucidation of a growth stimulant for arbuscular mycorrhizal fungus from Laminaria japonica Areschoug, *Journal of Applied Phycology*, 18(6), pp. 795-800.

Laghari, M., Hu, Z., Mirjat, M. S., Xiao, B., Tagar, A. A. and Hu, M. (2016), Fast pyrolysis biochar from sawdust improves the quality of desert soils and enhances plant growth, *Journal of the Science of Food and Agriculture*, 96(1), pp. 199-206. DOI:10.1002/jsfa.7082.

Lehmann, J., Rillig, M. C., Thies, J., Masiello, C. A., Hockaday, W. C. and Crowley, D. (2011), Biochar effects on soil biota – A review, *Soil Biology & Biochemistry*, 43(9), pp. 1812-36. 19th International Symposium on Environmental Biogeochemistry. DOI:10.1016/j.soilbio.2011.04.022.

Lim, S. L., Wu, T. Y., Lim, P. N. and Shak, K. P. Y. (2015), The use of vermicompost in organic farming: Overview, effects on soil and economics, *Journal of the Science of Food and Agriculture*, 95(6), pp. 1143-56. DOI:10.1002/jsfa.6849.

Lima, J. D., Zanetti, S., Nomura, E. S., Fuzitani, E. J., Rozane, D. E. and Iori, P. (2016), Growth and yield of anthurium in response to sawdust mulching, *Ciência Rural*, 46, pp. 440-6.

Lisiecka, J., Knaflewski, M., Spizewski, T., Fraszczak, B., Kaluzewicz, A. and Krzesinski, W. (2011), The effect of animal protein hydrolysate on quantity and quality of strawberry daughter plants cv.'Elsanta', *Acta Scientiarum Polonorum, Hortorum Cultus*, 10, pp. 31-40.

Liu, M., Wang, T., Cheng, Y., Fu, Y., Zhang, P., Xie, M., Sun, M. and Yang, D. (2014), Peat and brown coal resources in China and its potential for developing potassium humate fertilizer, *Earth Science Frontiers*, 5, pp. 25.

Løes, A.-K., Bünemann, E. K., Cooper, J., Hörtenhuber, S., Magid, J., Oberson, A. and Möller, K. (2016), Nutrient supply to organic agriculture as governed by EU regulations and standards in six European countries, *Organic Agriculture*, 7, pp. 395-418. DOI:10.1007/s13165-016-0165-3.

Macias-Corral, M. A., Samani, Z. A., Hanson, A. T. and Funk, P. A. (2017), Co-digestion of agricultural and municipal waste to produce energy and soil amendment, *Waste Management & Research*, 35(9), pp. 991-6. DOI:10.1177/0734242X17715097.

Mahdavi, B. and Rahimi, A. (2013), Seed priming with chitosan improves the germination and growth performance of ajowan (*Carum copticum*) under salt stress, *EurAsian Journal of BioSciences*, 7, 69-76.

Malusà, E., Sas Paszt, L., Głuszek, S. and Ciesielska, J. (2014), Organic fertilizers to sustain soil fertility, in: Sinha, S., Pant, K., Bajpai, S., and Govil, J. N. (Eds), *Fertilizers Technology Vol. I: Synthesis*. Houston, TX: Studium Press LLC, pp. 256-81.

Marinou, E., Chrysargyris, A. and Tzortzakis, N. (2013), Use of sawdust, coco soil and pumice in hydroponically grown strawberry, *Plant, Soil and Environment*, 59(10), pp. 452-9.

Marwaha, S. S. and Kennedy, J. F. (1988), Whey–pollution problem and potential utilization, *International Journal of Food Science and Technology*, 23(4), pp. 323-36. DOI:10.1111/j.1365-2621.1988.tb00586.x.

Masny, A., Basak, A. and Żurawicz, E. (2004), Effects of foliar applications of Kelpak SL and Goëmar BM 86®; preparations on yield and fruit quality in two strawberry cultivars, *Journal of Fruit and Ornamental Plant Research*, 12, pp. 23-7.

Mazzoncini, M., Antichi, D., Tavarini, S., Silvestri, N., Lazzeri, L. and D'Avino, L. (2015), Effect of defatted oilseed meals applied as organic fertilizers on vegetable crop production and environmental impact, *Industrial Crops and Products*, 75(Part A), pp. 54–64. DOI:10.1016/j.indcrop.2015.04.061.

Mobini, M., Khoshgoftarmanesh, A. H. and Ghasemi, S. (2014), The effect of partial replacement of nitrate with arginine, histidine, and a mixture of amino acids extracted from blood powder on yield and nitrate accumulation in onion bulb, *Scientia Horticulturae*, 176, pp. 232–7. DOI:10.1016/j.scienta.2014.07.014.

Möller, K. (2015), Assessment of alternative phosphorus fertilizers for organic farming: Meat and bone meal, Improve-P Factsheet, Universität Hohenheim, ETH Zürich, FiBL, Bioforsk, Universität für Bodenkultur Wien, Newcastle University, University of Copenhagen, pp. 1–8.

Möller, K. and Müller, T. (2012), Effects of anaerobic digestion on digestate nutrient availability and crop growth: A review, *Engineering in Life Sciences*, 12(3), pp. 242–57. DOI:10.1002/elsc.201100085.

Mosa, W.-G., Paszt, L. S., Frąc, M., Trzciński, P., Przybył, M., Treder, W. and Klamkowski, K. (2016), The influence of biofertilization on the growth, yield and fruit quality of cv. Topaz apple trees, *Horticultural Science*, 43(3), pp. 105–11.

Moutia, J.-F. Y., Saumtally, S., Spaepen, S. and Vanderleyden, J. (2010), Plant growth promotion by *Azospirillum* sp. in sugarcane is influenced by genotype and drought stress, *Plant and Soil*, 337(1), pp. 233–42. DOI:10.1007/s11104-010-0519-7.

Nagaraja, H., Chennappa, G., Rakesh, S., Naik, M. K., Amaresh, Y. S. and Sreenivasa, M. Y. (2016), Antifungal activity of *Azotobacter nigricans* against trichothecene-producing *Fusarium* species associated with cereals, *Food Science and Biotechnology*, 25(4), pp. 1197–204. DOI:10.1007/s10068-016-0190-8.

Ngala, B. M., Valdes, Y., dos Santos, G., Perry, R. N. and Wesemael, W. M. L. (2016), Seaweed-based products from *Ecklonia maxima* and *Ascophyllum nodosum* as control agents for the root-knot nematodes *Meloidogyne chitwoodi* and *Meloidogyne hapla* on tomato plants, *Journal of Applied Phycology*, 28(3), pp. 2073–82. DOI:10.1007/s10811-015-0684-4.

Nustorova, M., Braikova, D., Gousterova, A., Vasileva-Tonkova, E. and Nedkov, P. (2006), Chemical, microbiological and plant analysis of soil fertilized with alkaline hydrolysate of sheep's wool waste, *World Journal of Microbiology and Biotechnology*, 22(4), pp. 383–90. DOI:10.1007/s11274-005-9045-9.

Oni, O. E. and Friedrich, M. W. (2017), Metal oxide reduction linked to anaerobic methane oxidation, *Trends in Microbiology*, 25(2), pp. 88–90. DOI:10.1016/j.tim.2016.12.001.

Öpik, M., Vanatoa, A., Vanatoa, E., Moora, M., Davison, J., Kalwij, J. M., Reier, Ü. and Zobel, M. (2010), The online database MaarjAM reveals global and ecosystemic distribution patterns in arbuscular mycorrhizal fungi (Glomeromycota), *New Phytologist*, 188(1), pp. 223–41. DOI:10.1111/j.1469-8137.2010.03334.x.

Ostrovskij, M. (2014), Testing HUMIN PLUS microfertilizer, *European Agrophysical Journal*, 1 (2), pp. 79–84.

Ozkan, S. and Ozkan, S. G. (2016), Investigation of humate extraction from lignites, *International Journal of Coal Preparation and Utilization*, 37(6), pp. 285–92. DOI:10.1 080/19392699.2016.1171761.

Panda, D., Pramanik, K. B. and Naya, R. (2012), Use of sea weed extracts as plant growth regulators for sustainable agriculture, *International Journal of Stress Management*, 3(3), pp. 404–11.

Pane, C., Celano, G., Villecco, D. and Zaccardelli, M. (2012), Control of *Botrytis cinerea*, *Alternaria alternata* and *Pyrenochaeta lycopersici* on tomato with whey compost-tea applications, *Crop Protection*, 38, pp. 80–6. DOI:10.1016/j.cropro.2012.03.012.

Papenfus, H. B., Kulkarni, M. G., Stirk, W. A., Finnie, J. F. and Van Staden, J. (2013), Effect of a commercial seaweed extract (Kelpak®) and polyamines on nutrient-deprived (N, P and K) okra seedlings, *Scientia Horticulturae*, 151, pp. 142–6. DOI:10.1016/j.scienta.2012.12.022.

Paredes, C., Medina, E., Bustamante, M. A. and Moral, R. (2016), Effects of spent mushroom substrates and inorganic fertilizer on the characteristics of a calcareous clayey-loam soil and lettuce production, *Soil Use and Management*, 32(4), pp. 487–94. DOI:10.1111/sum.12304.

Parvage, M., Ulén, B., Eriksson, J., Strock, J. and Kirchmann, H. (2013), Phosphorus availability in soils amended with wheat residue char, *Biology and Fertility of Soils*, 49(2), pp. 245–50. DOI:10.1007/s00374-012-0746-6.

Peterson, A. E., Walker, W. G. and Watson, K. S. (1979), Effect of whey applications on chemical properties of soils and crops, *Journal of Agricultural and Food Chemistry*, 27(4), pp. 654–8. DOI:10.1021/jf60224a064.

Qian, P., Schoenau, J. and Urton, R. (2011), Effect of soil amendment with thin stillage and glycerol on plant growth and soil properties, *Journal of Plant Nutrition*, 34(14), pp. 2206–21. DOI:10.1080/01904167.2011.618579.

Qian, S., Ding, W., Li, Y., Liu, G., Sun, J. and Ding, Q. (2015), Characterization of humic acids derived from Leonardite using a solid-state NMR spectroscopy and effects of humic acids on growth and nutrient uptake of snap bean, *Chemical Speciation & Bioavailability*, 27 (4), pp. 156–61. DOI:10.1080/09542299.2015.1118361.

Raghu, K. and MacRae, I. C. (1966), Occurrence of phosphate-dissolving micro-organisms in the rhizosphere of rice plants and in submerged soils, *Journal of Applied Bacteriology*, 29(3), pp. 582–6. DOI:10.1111/j.1365-2672.1966.tb03511.x.

Rashid, A., Mir, M. R. and Hakeem, K. R. (2016a), Biofertilizer use for sustainable agricultural production, in: Hakeem, K. R., Akhtar, M. S., and Abdullah, S. N. A. (Eds), *Plant, Soil and Microbes: Volume 1: Implications in Crop Science*. Cham, Switzerland: Springer International Publishing, pp. 163–80.

Rashid, M. I., Mujawar, L. H., Shahzad, T., Almeelbi, T., Ismail, I. M. I. and Oves, M. (2016b), Bacteria and fungi can contribute to nutrients bioavailability and aggregate formation in degraded soils, *Microbiological Research*, 183, pp. 26–41. DOI:10.1016/j.micres.2015.11.007.

Reed, E. Y., Chadwick, D. R., Hill, P. W. and Jones, D. L. (2017), Critical comparison of the impact of biochar and wood ash on soil organic matter cycling and grassland productivity, *Soil Biology and Biochemistry*, 110, pp. 134–42. DOI:10.1016/j.soilbio.2017.03.012.

Robbins, C. W. and Lehrsch, G. A. (1992), Effects of acidic cottage cheese whey on chemical and physical properties of a sodic soil, *Arid Soil Research and Rehabilitation*, 6 (2), pp. 127–34. DOI:10.1080/15324989209381305.

Roger, P. A. and Ladha, J. K. (1992), Biological N2 Fixation in wetland rice fields: Estimation and contribution to nitrogen balance, *Plant and Soil*, 141(1), pp. 41–55. DOI:10.1007/BF00011309.

Rouphael, Y., Franken, P., Schneider, C., Schwarz, D., Giovannetti, M., Agnolucci, M., Pascale, S. D., Bonini, P. and Colla, G. (2015), Arbuscular mycorrhizal fungi act as biostimulants in horticultural crops, *Biostimulants in Horticulture*, 196, pp. 91-108. DOI:10.1016/j.scienta.2015.09.002.

Roy, S., Barman, S., Chakraborty, U. and Chakraborty, B. (2015), Evaluation of spent mushroom substrate as biofertilizer for growth improvement of *Capsicum annuum* L, *Journal of Applied Biology and Biotechnology*, 3(03), pp. 22-7.

Roychowdhury, D., Paul, M. and KumarBanerjee, S. (2015), Isolation identification and characterization of phosphate solubilising bacteria from soil and the production of biofertilizer, *International Journal of Current Microbiology and Applied Sciences*, 4 (11), pp. 808-15.

Rozpara, E., Pąśko, M., Bielicki, P. and Sas Paszt, L. (2014), Influence of various bio-fertilizers on the growth and fruiting of 'Ariwa' apple trees growing in an organic orchard, *Journal of Research and Applications in Agricultural Engineering*, 59(4), pp. 65-8.

Rutkowska, B., Szulc, W., Sosulski, T. and Stepien, W. (2014), Soil micronutrient availability to crops affected by long-term inorganic and organic fertilizer applications, *Plant, Soil and Environment*, 60(5), pp. 198-203.

Saito, B. and Seckler, M. M. (2014), Alkaline extraction of humic substances from peat applied to organic-mineral fertilizer production, *Brazilian Journal of Chemical Engineering*, 31, pp. 675-82.

Sajbrt, V., Rosol, M. and Ditl, P. (2010), A comparison of distillery stillage disposal methods, *Acta Polytechnica*, 50(2), pp. 63-9.

Salama, A. S., El-Sayed, O. M. and El Gammal, O. (2014), Effect of Effective Microorganisms (EM) and potassium sulphate on productivity and fruit quality of 'Hayany' date palm grown under salinity stress, *Journal of Agriculture and Veterinary Sciences*, 7, pp. 90-9.

Salantur, A., Ozturk, A. and Akten, S. (2006), Growth and yield response of spring wheat (*Triticum aestivum* L.) to inoculation with rhizobacteria, *Plant, Soil and Environment*, 52(3), pp. 111-18.

Sanderson, K. R. and Cutcliffe, J. A. (1991), Effect of sawdust mulch on yields of select clones of lowbush blueberry, *Canadian Journal of Plant Science*, 71(4), pp. 1263-6. DOI:10.4141/cjps91-175.

Sas Paszt, L., Sumorok, B., Malusá, E., Głuszek, S. and Derkowska, E. (2011), The influence of bioproducts on root growth and mycorrhizal occurrence in the rhizosphere of strawberry plants 'Elsanta', *Journal of Fruit and Ornamental Plant Research*, 19(1), pp. 13-34.

Sas Paszt, L., Malusá, E., Sumorok, B., Canfora, L., Derkowska, E. and Głuszek, S. (2015), The influence of bioproducts on mycorrhizal occurrence and diversity in the rhizosphere of strawberry plants under controlled conditions, *Advances in Microbiology*, 5(1), pp. 40-53.

Sasirekha, B. and Srividya, S. (2016), Siderophore production by *Pseudomonas aeruginosa* FP6, a biocontrol strain for *Rhizoctonia solani* and *Colletotrichum gloeosporioides* causing diseases in chilli, *Agriculture and Natural Resources*, 50(4), pp. 250-6. DOI:10.1016/j.anres.2016.02.003.

Schäfer, T., Hanke, M.-V., Flachowsky, H., König, S., Peil, A., Kaldorf, M., Polle, A. and Buscot, F. (2012), Chitinase activities, scab resistance, mycorrhization rates and biomass of own-rooted and grafted transgenic apple, *Genetics and Molecular Biology*, 35(2), pp. 466-73. DOI:10.1590/S1415-47572012000300014.

Schlegel, A. J., Assefa, Y., Bond, H. D., Haag, L. A. and Stone, L. R. (2017), Changes in soil nutrients after 10 years of cattle manure and swine effluent application, *Soil and Tillage Research*, 172, pp. 48–58. DOI:10.1016/j.still.2017.05.004.

Shaheen, S., Khan, M., Khan, M. J., Jilani, S., Bibi, Z., Munir, M. and Kiran, M. (2017), Effective Microorganisms (EM) co-applied with organic wastes and NPK stimulate the growth, yield and quality of spinach (*Spinacia oleracea* L.), *Sarhad Journal of Agriculture*, 33 (1), pp. 30–41.

Sharp, G. R. (2013), A review of the applications of chitin and its derivatives in agriculture to modify plant- microbial interactions and improve crop yields, *Agronomy*, 3(4), pp. 778–93. DOI:10.3390/agronomy3040757.

Sharratt, W. J., Peterson, A. E. and Calbert, H. E. (1959), Whey as a source of plant nutrients and its effect on the soil, *Journal of Dairy Science*, 42(7), pp. 1126–31. DOI:10.3168/jds.S0022-0302(59)90705-2.

Siebers, N., Godlinski, F. and Leinweber, P. (2014), Bone char as phosphorus fertilizer involved in cadmium immobilization in lettuce, wheat, and potato cropping, *Journal of Plant Nutrition and Soil Science*, 177(1), pp. 75–83. DOI:10.1002/jpln.201300113.

Sim, E. Y. S. and Wu, T. Y. (2010), The potential reuse of biodegradable municipal solid wastes (MSW) as feedstocks in vermicomposting, *Journal of the Science of Food and Agriculture*, 90(13), pp. 2153–62. DOI:10.1002/jsfa.4127.

Singh, R., Gupta, R. K., Patil, R. T., Sharma, R. R., Asrey, R., Kumar, A. and Jangra, K. K. (2010), Sequential foliar application of vermicompost leachates improves marketable fruit yield and quality of strawberry (*Fragaria × ananassa* Duch.), *Scientia Horticulturae*, 124(1), pp. 34–9. DOI:10.1016/j.scienta.2009.12.002.

Singh, P., Singh, M. and Tripathi, B. (2013), Glomalin: An arbuscular mycorrhizal fungal soil protein, *Protoplasma*, 250(3), pp. 663–9. DOI:10.1007/s00709-012-0453-z.

Singh, A. K., Beer K., and Kumar Pal, A. K. (2015), Effect of vermicompost and biofertilizers on strawberry I: Growth, flowering and yield, *Annals of Plant and Soil Research*, 17, pp. 196–9.

Singh, A., Kaushik, M. S., Srivastava, M., Tiwari, D. N. and Mishra, A. K. (2016), Siderophore mediated attenuation of cadmium toxicity by paddy field cyanobacterium *Anabaena oryzae*, *Algal Research*, 16, pp. 63–8. DOI:10.1016/j.algal.2016.02.030.

Sohi, S. P., Krull, E., Lopez-Capel, E. and Bol, R. (2010), A review of biochar and its use and function in soil, *Advances in Agronomy*, 105, pp. 47–82. DOI:10.1016/S0065-2113(10)05002-9.

Souza, E. M., Chubatsu, L. S., Huergo, L. F., Monteiro, R., Camilios-Neto, D., Wassem, R. and de Oliveira Pedrosa, F. (2014), Use of nitrogen-fixing bacteria to improve agricultural productivity, *BMC Proceedings*, 8(4), pp. O23. DOI:10.1186/1753-6561-8-S4-O23.

Spokas, K. A. (2010), Review of the stability of biochar in soils: Predictability of O:C molar ratios, *Carbon Management*, 1(2), pp. 289–303. DOI:10.4155/cmt.10.32.

Spokas, K. A., Koskinen, W. C., Baker, J. M. and Reicosky, D. C. (2009), Impacts of woodchip biochar additions on greenhouse gas production and sorption/degradation of two herbicides in a Minnesota soil, *Chemosphere*, 77(4), pp. 574–81. DOI:10.1016/j.chemosphere.2009.06.053.

Spokas, K. A., Cantrell, K. B., Novak, J. M., Archer, D. W., Ippolito, J. A., Collins, H. P., Boateng, A. A., Lima, I. M., Lamb, M. C., McAloon, A. J., Lentz, R. D. and Nichols, K. A. (2012a), Biochar: A synthesis of its agronomic impact beyond carbon sequestration, *Journal of Environmental Quality*, 41(4), pp. 973–89.

Spokas, K., Novak, J. and Venterea, R. (2012b), Biochar's role as an alternative N-fertilizer: Ammonia capture, *Plant and Soil*, 350(1-2), pp. 35-42. DOI:10.1007/s11104-011-0930-8.

Stadnik, M. J. and Freitas, M. B. de (2014), Algal polysaccharides as source of plant resistance inducers, *Tropical Plant Pathology*, 39(2), pp. 111-18.

Stefaniuk, M., Bartmiński, P., Różyło, K., Dębicki, R. and Oleszczuk, P. (2015), Ecotoxicological assessment of residues from different biogas production plants used as fertilizer for soil, *Journal of Hazardous Materials*, 298, pp. 195-202. DOI:10.1016/j.jhazmat.2015.05.026.

Stefaniuk, M., Oleszczuk, P. and Bartmiński, P. (2016), Chemical and ecotoxicological evaluation of biochar produced from residues of biogas production, *Journal of Hazardous Materials*, 318, pp. 417-24. DOI:10.1016/j.jhazmat.2016.06.013.

Steiner, C., Das, K. C., Garcia, M., Förster, B. and Zech, W. (2008), Charcoal and smoke extract stimulate the soil microbial community in a highly weathered xanthic Ferralsol, *Pedobiologia*, 51(5-6), pp. 359-66. DOI:10.1016/j.pedobi.2007.08.002.

Steiner, C., Garcia, M. and Zech, W. (2009), Effects of charcoal as slow release nutrient carrier on N-P-K dynamics and soil microbial population: Pot experiments with Ferralsol substrate, in: Woods, W., Teixeira, W., Lehmann, J., Steiner, C., WinklerPrins, A., and Rebellato, L. (Eds), *Amazonian Dark Earths: Wim Sombroek's Vision*. Dordrecht, the Netherlands: Springer Netherlands, pp. 325-38.

Stępień, W., Malusà, E., Paszt, L. S., Renzi, G. and Ciesielska, J. (2012), Effect of brown coal- based composts produced with the use of white rot fungi on the growth and yield of strawberry plants, in: Ecofruit. *15th International Conference on Organic Fruit-Growing. Proceedings for the Conference*. Hohenheim, Germany: Fördergemeinschaft Ökologischer Obstbau eV (FÖKO), pp. 302-6.

Stirk, W. A. and Staden, J. (1996), Comparison of cytokinin- and auxin-like activity in some commercially used seaweed extracts, *Journal of Applied Phycology*, 8(6), pp. 503-8. DOI:10.1007/BF02186328.

Struszczyk, H., Pospieszny H., and Kotliński S. (1989), Some new application of chitosan, in: Skjoak-Bræk G., Anthonsen T., Sandford P. (Eds), *Chitin and Chitosan*, Elsevier Science, London. pp. 733-42.

Sumbul, A., Rizvi, R., Mahmood, I. and Ansari, R. A. (2015), Oil-cake amendments: Useful tools for the management of phytonematodes, *Asian Journal of Plant Pathology*, 9(3), pp. 91-111.

Sun, Q., Ding, W., Yang, Y., Sun, J. and Ding, Q. (2016), Humic acids derived from leonardite-affected growth and nutrient uptake of corn seedlings, *Communications in Soil Science and Plant Analysis*, 47(10), pp. 1275-82. DOI:10.1080/00103624.2016.1178767.

Szajdak, L. W. (2016), Phytohormone in peats, sapropels, and peat substrates, in: Szajdak, L.W. (Ed.), *Bioactive Compounds in Agricultural Soils*. Cham: Springer International Publishing, pp. 247-72.

Szpak, P., Longstaffe, F. J., Millaire, J.-F. and White, C. D. (2012), Stable isotope biogeochemistry of seabird guano fertilization: Results from growth chamber studies with maize (*Zea mays*), *PLoS ONE*, 7 (3), pp. e33741 (1-16). DOI:10.1371/journal.pone.0033741.

Taktek, S., Trépanier, M., Servin, P. M., St-Arnaud, M., Piché, Y., Fortin, J.-A. and Antoun, H. (2015), Trapping of phosphate solubilizing bacteria on hyphae of the arbuscular mycorrhizal fungus Rhizophagus irregularis DAOM 197198, *Soil Biology and Biochemistry*, 90, pp. 1–9. DOI:10.1016/j.soilbio.2015.07.016.

Tejeda-Agredano, M.-C., Mayer, P. and Ortega-Calvo, J.-J. (2014), The effect of humic acids on biodegradation of polycyclic aromatic hydrocarbons depends on the exposure regime, *Environmental Pollution*, 184, pp. 435–42. DOI:10.1016/j.envpol.2013.09.031.

Thorsen, M., Woodward, S. and McKenzie, B. (2010), Kelp (*Laminaria digitata*) increases germination and affects rooting and plant vigour in crops and native plants from an arable grassland in the Outer Hebrides, Scotland, *Journal of Coastal Conservation*, 14(3), pp. 239–47. DOI:10.1007/s11852-010-0091-6.

Toselli, M., Sorrenti, G., Marangoni, B., Innocenti, A., Baldi, E., Marcolini, G. and Quartieri, M. (2013), Effect of organic fertilization on soil fertility, tree nutritional status and nutrient removal of mature nectarine trees, *Acta Horticulturae*, 1001, pp. 303–10.

Tretjakova, R., Grebeža, J. and Martinovs, A. (2015), Research into biological characteristics of dried sapropel, in: *Environment. Technology, Resources*. Rezekne, Latvia, 1, pp. 223–7.

Troy, S. M., Nolan, T., Kwapinski, W., Leahy, J. J., Healy, M. G. and Lawlor, P. G. (2012), Effect of sawdust addition on composting of separated raw and anaerobically digested pig manure, *Journal of Environmental Management*, 111, pp. 70–7. DOI:10.1016/j.jenvman.2012.06.035.

Vaneeckhaute, C., Meers, E., Michels, E., Buysse, J. and Tack, F. M. G. (2013), Ecological and economic benefits of the application of bio-based mineral fertilizers in modern agriculture, *Biomass and Bioenergy*, 49, pp. 239–48. DOI:10.1016/j.biombioe.2012.12.036.

Vanek, S. J., Thies, J., Wang, B., Hanley, K. and Lehmann, J. (2016), Pore-size and water activity effects on survival of Rhizobium tropici in biochar inoculant carriers, *Journal of Microbial & Biochemical Technology*, 8, pp. 296–306.

Vasconcellos, R. L. F., Bonfim, J. A., Baretta, D. and Cardoso, E. J. B. N. (2016), Arbuscular mycorrhizal fungi and glomalin-related soil protein as potential indicators of soil quality in a recuperation gradient of the atlantic forest in Brazil, *Land Degradation & Development*, 27(2), pp. 325–34. DOI:10.1002/ldr.2228.

Velázquez, E., Carro, L., Flores-Félix, J. D., Martínez-Hidalgo, P., Menéndez, E., Ramírez-Bahena, M.-H., Mulas, R., González-Andrés, F., Martínez-Molina, E. and Peix, A. (2017), The legume nodule microbiome: A source of plant growth- promoting bacteria, in: Kumar, V., Kumar, M., Sharma, S., and Prasad, R. (Eds), *Probiotics and Plant Health*. Singapore: Springer Singapore, pp. 41–70.

Vessey, J. K. (2003), Plant growth promoting rhizobacteria as biofertilizers, *Plant and Soil*, 255(2), pp. 571–86. DOI:10.1023/A:1026037216893.

Vicario, J. C., Primo, E. D., Dardanelli, M. S. and Giordano, W. (2016), Promotion of peanut growth by co- inoculation with selected strains of Bradyrhizobium and Azospirillum, *Journal of Plant Growth Regulation*, 35(2), pp. 413–19. DOI:10.1007/s00344-015-9547-0.

Walpola, B. C. and Yoon, M.-H. (2012), Prospectus of phosphate solubilizing microorganisms and phosphorus availability in agricultural soils: A review, *African Journal of Microbiology Research*, 6(37), pp. 6600–5. DOI:10.5897/AJMR12.889

Wani, S. P. and Lee, K. K. (2002), Population dynamics of nitrogen fixing bacteria associated with pearl millet (*P. americanum* L.), in: *Biotechnology of Nitrogen Fixation in the Tropics*. University of Pertanian, Malaysia, pp. 21–30.

Wani, F. S., Latief Ahmad, T. A. and Mushtaq, A. (2015), Role of microorganisms in nutrient mobilization and soil health-A review, *Journal of Pure and Applied Microbiology*, 9(2), pp. 1401-10.

Wilkie, A. C., Riedesel, K. J. and Owens, J. M. (2000), Stillage characterization and anaerobic treatment of ethanol stillage from conventional and cellulosic feedstocks, *Biomass and Bioenergy*, 19(2), pp. 63-102. DOI:10.1016/S0961-9534(00)00017-9.

Yang, X., Wang, X., Wang, K., Su, L., Li, H., Li, R. and Shen, Q. (2015a), The nematicidal effect of Camellia seed cake on root-knot nematode Meloidogyne javanica of banana, *PLoS ONE*, 10 (4), pp. e0119700 (1-18). DOI:10.1371/journal.pone.0119700.

Yang, L., Zhao, F., Chang, Q., Li, T. and Li, F. (2015b), Effects of vermicomposts on tomato yield and quality and soil fertility in greenhouse under different soil water regimes, *Agricultural Water Management*, 160, pp. 98-105. DOI:10.1016/j.agwat.2015.07.002.

Yooyongwech, S., Samphumphuang, T., Tisarum, R., Theerawitaya, C. and Cha-um, S. (2016), Arbuscular mycorrhizal fungi (AMF) improved water deficit tolerance in two different sweet potato genotypes involves osmotic adjustments via soluble sugar and free proline, *Scientia Horticulturae*, 198, pp. 107-17. DOI:10.1016/j.scienta.2015.11.002.

Yu, T., Wang, L., Yin, Y., Wang, Y. and Zheng, X. (2008), Effect of chitin on the antagonistic activity of *Cryptococcus laurentii* against *Penicillium expansum* in pear fruit, *International Journal of Food Microbiology*, 122(1-2), pp. 44-8. DOI:10.1016/j.ijfoodmicro.2007.11.059

Yu, G., Ran, W. and Shen, Q. (2016), Compost process and organic fertilizers application in China, in: Larramendy, M. L. and Soloneski, S. (Eds), *Organic Fertilizers - From Basic Concepts to Applied Outcomes*. Rijeka: InTech, pp. 1-24.

Yusuf, R., Kristiansen, P. and Warwick, N. (2016), Effect of two seaweed products on radish (*Raphanus sativus*) growth under greenhouse conditions, *AGROLAND: The Agricultural Sciences Journal*, 2(1), pp. 1-7.

Zeng, D., Luo, X. and Tu, R. (2012), Application of bioactive coatings based on chitosan for soybean seed protection, *International Journal of Carbohydrate Chemistry*, 2012, pp. 104565 (1-5).

Zhang, H., Ding, W., He, X., Yu, H., Fan, J. and Liu, D. (2014), Influence of 20-year organic and inorganic fertilization on organic carbon accumulation and microbial community structure of aggregates in an intensively cultivated sandy loam soil, *PLoS ONE*, 9(3), pp. e92733. DOI:10.1371/journal.pone.0092733.

Zhao, R., Guo, W., Bi, N., Guo, J., Wang, L., Zhao, J. and Zhang, J. (2015), Arbuscular mycorrhizal fungi affect the growth, nutrient uptake and water status of maize (*Zea mays* L.) grown in two types of coal mine spoils under drought stress, *Applied Soil Ecology*, 88, pp. 41-9. DOI:10.1016/j.apsoil.2014.11.016.

Zhen, Z., Liu, H., Wang, N., Guo, L., Meng, J., Ding, N., Wu, G. and Jiang, G. (2014), Effects of manure compost application on soil microbial community diversity and soil microenvironments in a temperate cropland in China, *PLoS ONE*, 9(10), pp. e108555. DOI:10.1371/journal.pone.0108555.

Chapter 2

Assessing the effects of compost on soil health

Cristina Lazcano, University of California-Davis, USA; Charlotte Decock, California Polytechnic State University, USA; Connie T. F. Wong, University of California-Davis, USA; and Kamille Garcia-Brucher, California Polytechnic State University, USA

1 Introduction

2 Why compost?

3 Effects of compost on soil nutrient cycling

4 Effects of compost on soil hydraulic properties

5 Effect of compost on crop productivity

6 Effects of compost on soil biodiversity

7 Effects of compost on environmental quality

8 The use of compost to improve soil health in annual crops: a case study with strawberries

9 The use of compost to improve soil health, sequester carbon and reduce greenhouse gas emissions in perennial crops: a case study in a Mediterranean vineyard

10 Conclusion

11 Where to look for further information

12 References

1 Introduction

Returning organic waste materials to the soil in the form of compost or other organic amendments has been identified as one of the key strategies to promote soil health and slow down or reverse rapid degradation of agricultural lands (Karlen et al., 2019). Organic amendments or organic fertilizers are naturally occurring organic materials from animal or plant origin that can be found in different degrees of decomposition or biochemical stability and support plant growth by improving soil physical properties and/or providing essential plant nutrients. Examples of organic amendments include animal manures, compost, vermicompost, biochar, or crop residues, among others.

http://dx.doi.org/10.19103/AS.2021.0094.06

These amendments provide diverse food sources for soil biota and part of the organic material becomes incorporated into soil organic matter (SOM) upon decomposition (Lehmann and Kleber, 2015). Buildup of SOM, changes in the soil microbiome, and beneficial effects of decomposition by-products such as plant nutrients and organic compounds ultimately determine the outcome of organic amendments for soil health (Diacono and Montemurro, 2010; Mehta et al., 2014). Nevertheless, given the diversity of organic amendments, both in terms of chemical composition and in biochemical stability, the benefits for soil health are also highly variable in magnitude. Important properties such as the total plant macro- and micronutrient content, carbon to nitrogen ratio (C:N), salinity, and presence of pathogenic microorganisms or heavy metals can significantly affect the outcomes for soil health. Factors such as the initial feedstock, composting process, or age can significantly affect the physicochemical and biological properties of the finished compost (Hargreaves et al., 2008).

When management strategies to improve soil health such as compost application are adopted, it is therefore pertinent to track management-induced improvements in soil health over time, taking into account the specific properties of the compost used. Standard soil testing for agronomic management is heavily focused on chemical soil health indicators such as soil nutrient concentrations and pH, greatly ignoring biological soil health and the role of soil physical properties in promoting a suitable habitat for soil biota and root growth (Lazcano et al., 2021a). The concept of soil health, on the other hand, not only incorporates soil chemical parameters but integrates them with physical and biological soil parameters, emphasizing the functioning of soils as ecosystems (Pawlett et al., 2021). Various indicators to assess soil health have been proposed over the past few decades. These include measurable physical, chemical, and biological soil properties that are indicative of the functions and ecosystem services provided by healthy soils (Fig. 1). Nevertheless, poor standardization and the lack of robust interpretation schemes limit the capacity of soil health indicators for predicting and assessing soil threats, soil functioning, or provisioning of ecosystem services (Bünemann et al., 2018). In the light of an increasing need to assess and quantify the impact of management practices that promote soil health, recent studies have aimed to standardize the measurement and interpretability of soil health indicators (Moebius-Clune, 2016; Norris et al., 2020; Nunes et al., 2020; Rinot et al., 2019). Given the multifunctional nature of soil health, no single soil health indicator would accurately reflect soil health. Additionally, narrowing down soil health assessments to a set of indicators that comprehensively capture and accurately rate the many functions healthy soils provide, from suppressing plant disease and supplying plant nutrients to filtering water and sequestering carbon, is a challenging task (Wade et al., 2022).

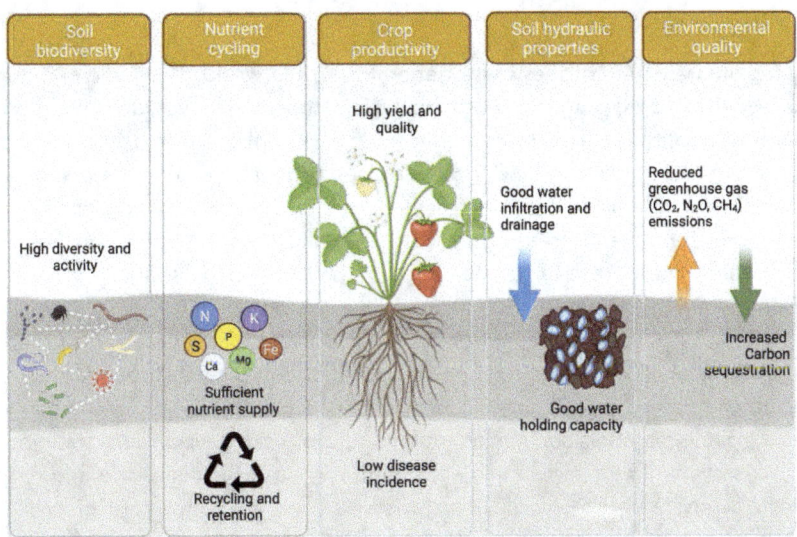

Figure 1 Overview of the functions performed by a healthy soil. Figure created with Biorender.com.

A great wealth of field and incubation experiments carried out in the last decades has shown that the addition of compost to soils has well-known short- and long-term, direct, and indirect benefits to soil biological, chemical, and physical properties that support ecosystem functioning and soil health (Diacono and Montemurro, 2010; Martinez-Blanco et al., 2013a,b). Nevertheless, these studies show substantial variability in the observed effects. This lack of clarity may result in the decrease of grower adoption rates and make the precision management of this soil organic amendment very difficult. As growers around the world move towards more sustainable soil management practices and embrace the principles of the circular economy, it is expected that the use of compost will increase and become more widespread across production systems. Thus, there is a need for a critical evaluation of the potential of compost to improve soil health, what are the best indicators to evaluate these benefits, and what are the expected ranges of variability.

This chapter provides an overview of the current literature evaluating the effects of compost on soil health in the short-, medium-, and long-term. By looking at a wide range of soil health indicators and crops, we aim to identify those indicators that are consistently responsive to compost application, as well as to characterize potential sources of variability. Finally, we identify potential risks of the use of compost for environmental quality, pointing at future research needs. Two case studies illustrate examples of current research evaluating the potential tradeoffs between benefit and risks in the use of compost in perennial and annual crops.

2 Why compost?

Approximately 44% of the waste generated in the world is organic in nature (Kaza et al., 2018). Organic waste materials constitute a valuable source of plant nutrients that, if used as fertilizers, can contribute to closing the loop of nutrients in agriculture and increasing soil organic matter with benefits for soil physical, chemical, and biological properties (Diacono and Montemurro, 2010). Furthermore, increasing soil organic matter contributes to increasing soil C and therefore is an appealing pathway for soil C sequestration and climate change mitigation. Finally, recycling of these materials eliminates the emissions of pollutants (i.e. greenhouse gasses, heavy metals) and nutrients associated with landfilling, thus having additional benefits for environmental health.

Organic materials such as animal manure or crop residues are traditional sources of plant nutrients which are commonly used in small-scale and crop-livestock integrated systems. These materials are also important in organic production systems where crop nutrient inputs rely heavily on organic fertilizers (Ramos et al., 2019). Production of animal manures has experienced a steady increase worldwide due to the increase in the consumption of livestock and poultry (Bai et al., 2018; Chadwick et al., 2015), which could help satisfy the needs for fertilizer products in agricultural systems. For example, only in 2011, 120 Tg of manure N were excreted, an amount which was similar or larger than the total synthetic fertilizer N requirements worldwide (Liu et al., 2017). Nevertheless, the reintegration of animal manures into cropping systems is challenging due to the presence of enteric microorganisms in animal manures such as *Escherichia coli* and *Salmonella* sp. (Tran et al., 2020). The spread of animal manures is also a source of antibiotic resistance in agroecosystems (Wind et al., 2021). Furthermore, mismanagement of animal manures can increase the contents of salts in the soil and the availability of labile forms of N and C that can induce nitrification and denitrification processes, leading to the loss of large amounts of N (Charles et al., 2017; Lazcano et al., 2021b). Similar concerns are associated with the agricultural use of certain plant-based organic waste materials. For example, grape pomace is a nutrient-rich solid waste material produced in large amounts in wine-growing regions of the world, which consists of the seeds, skins, and residual stalks of pressed grapes (Oliveira and Duarte, 2016). Due to its high phenolic content, low pH, high C:N, and large concentration of salts, grape pomace is highly toxic for plant growth. Similarly, the application of sewage sludge or municipal solid waste in agriculture poses several environmental and agronomic challenges due to the presence of human pathogens, plastics, and contaminants of different nature including heavy metals or solvents (Langdon et al., 2019; McBride, 1995; Petrie et al., 2015).

Composting involves the accelerated aerobic decomposition of organic waste materials by microorganisms, under controlled conditions of temperature and humidity (Debertoldi et al., 1983). The breakdown of labile organic compounds allows for the biological stabilization of the organic waste material decreasing the amount of easily available C and N and therefore the risks of environmental pollution after application to the soil compared to uncomposted materials (Charles et al., 2017; Li et al., 2016; Oudart et al., 2012). During composting, the organic waste typically undergoes a characteristic thermophilic stage, resulting from the intense microbial activity, which allows for the elimination of pathogenic microorganisms (Mc Carthy et al., 2011). Subsequently, during the maturation stage, decomposition rates decrease, together with compost temperature. Composting has demonstrated to be a robust, low-cost technology to efficiently stabilize and valorize several types of organic waste materials including animal manures (Bernal et al., 2009; Gómez-Brandón et al., 2008), sewage sludge (Lü et al., 2021), municipal solid waste (Kumar, 2011), anaerobic digestate (Bustamante et al., 2012); grape pomace (Ruggieri et al., 2009), paper pulp (Jain et al., 2018; Sesay et al., 1997), and other agricultural and food processing waste materials. The highest process efficiency and compost quality is achieved through co-composting which consists in the mix of different feedstocks with different N contents and particle sizes, to obtain a material with optimal physicochemical properties that support microbial activity and decomposition (Bustamante et al., 2008; Hazarika and Khwairakpam, 2018). Thus, most commercial composts are produced with a mix of plant and animal organic waste materials (Bernal et al., 2017).

Composting of organic waste and subsequent application to agricultural land is arguably one of the most impactful strategies to restore and maintain soil health due to the accessibility and low cost of composting technologies, particularly at the farm scale (Colvero et al., 2020; Pergola et al., 2018). Different composting systems exist, ranging from closed, automated in-vessel reactor systems to open static piles (Bernal et al., 2017). Closed in-vessel systems provide forced aeration and controlled ventilation, therefore accelerating decomposition and reducing the losses of N through leaching of nitrates or through the release of gases (NH_3 or N_2O). On the other hand, static piles typically involve minimal external intervention and rely on the presence of a critical mass that will result in the self-heating of the waste. Windrow composting involves a higher manipulation of the pile using specialized tractors to mix and aerate the waste material. The output of gases and leachates in windrow and static pile composting systems is not controlled, which results in larger losses of N compared to in-vessel reactors (Liu et al., 2020; Makan and Fadili, 2020). While closed composting systems are certainly more efficient and environmentally friendly, they are also less affordable. Thus, static piles or windrow composting may be more feasible and accessible for small commercial or on-farm operations. The simplicity, robustness and affordability of composting makes it an attractive strategy for management of organic waste at different scales, from home composting to integration of several waste streams at the regional level. Pergola et al. (2018) determined that on-farm composting was the most environmentally friendly strategy for the conversion of organic solid waste into compost in southern Italy. In China, about 76% of the animal manure produced in intensive livestock farms is composted on-site, while 9% is composted off-farm (Chadwick et al., 2015). In urban regions, home composting is an effective strategy in diverting organic waste away from landfilling, as shown by Pai et al. (2019) in a study carried out in Chicago (USA).

Compost, the final product of composting, is a stabilized and decomposed organic material which is compatible and beneficial to plant growth (Bernal et al., 2009; Insam and de Bertoldi, 2007). Given the broad definition of compost and the large range of feedstock materials that can be used in composting, there is also large variability in compost quality and physicochemical properties (Epelde et al., 2018). Most regions have established quality standards for commercially produced compost that regulate parameters such as odors, microbial stability, nutrient contents, contaminants (e.g. heavy metals, PAHs), percentage of impurities (plastic, glass, and metal), pH, soluble salts, moisture content, organic matter content, pathogens (*Salmonella* and fecal coliforms), and maturity (determined in seed germination assays) (Bernal et al., 2017; Bläsing and Amelung, 2018). The quality of the initial feedstock has a great impact on the quality of the final compost and, while many quality parameters can be regulated through adequate management of the composting process (Lannan et al., 2013), some others such as impurities or contaminants are more challenging. Typically, compost obtained from the organic fraction of municipal solid waste tends to be of lower quality than compost obtained from manure and/or green waste, due to a higher percentage of salts (contained in MSW and further increased through mineralization), impurities, and contaminants which are not eliminated through composting (de Araújo et al., 2010; Hargreaves et al., 2008). Thus, many countries such as Spain have strict quality criteria that regulates the end use of municipal solid waste compost defining limits for heavy metal concentration and pathogens (Bernal et al., 2017; Siles-Castellano et al., 2020). Nevertheless, other countries have looser requirements for contaminants and pathogens, and several researchers are calling for a better harmonization of criteria at the international level (Bernal et al., 2017). For this review, we will assume that compost meets the highest quality standards to be used as a soil amendment or fertilizer in agricultural land.

3 Effects of compost on soil nutrient cycling

3.1 Soil macro- and micronutrient content

Compost, as stabilized organic matter, is generally a nutrient-rich material which contains all essential plant macro- and micronutrients (Diacono and Montemurro, 2010; Duong et al., 2013; Martínez-Blanco et al., 2013b). Nevertheless, both the total nutrient content and nutrient availability depend largely on the compost feedstock(s) and composting process. The content of available or potentially plant available N, the most limiting nutrient for plant growth, can range from 4 g kg^{-1} dry matter to 120 g kg^{-1} dry matter, depending on the compost feedstock (Bernal et al., 2017). Total P contents range from 0.4 g kg^{-1} dry matter to 23 g kg^{-1} dry matter, and K contents range from 0.7 g kg^{-1} to 12 g kg^{-1} dry matter (Stoffella et al., 2001). Generally, N, P, and K contents tend to be lower in plant-based or compost produced from certain industrial feedstocks such as paper pulp, while compost produced from animal manures, on the other hand, typically have higher N and P concentrations (Chen et al., 2018; Habteweld et al., 2018). The concentration of micronutrients in compost is also highly variable; for example, Stoffella et al. (2001) and Siles-Castellano et al. (2020) reported that in municipal solid waste, compost samples collected in the US and Europe Fe content ranged between 9300 mg kg^{-1} dw and 16 400 mg kg^{-1} dw, B between 3 mg kg^{-1} dw and 60 mg kg^{-1} dw, Cu between 31 mg kg^{-1} dw and 1052 mg kg^{-1} dw, Mn between 400 mg kg^{-1} dw and 600 mg kg^{-1} dw, Mo 7.2 mg kg^{-1} dw, and Zn between 15 mg kg^{-1} dw and 1650 mg kg^{-1} dw. The composting process also decreases nutrient availability and therefore potential release to the soil as compared to raw organic waste materials (Chalk et al., 2013).

Compost can be used to replace synthetic fertilizers in agricultural systems partially or completely (Al-Suhaibani et al., 2020; Lazcano et al., 2013; Wu and Ma, 2015). Interestingly, due to the complex chemical composition of composts as compared to the most common NPK synthetic fertilizers, compost constitutes an excellent way of replenishing both plant macro- and micronutrients in soils. Ample evidence in the literature shows that compost application increases soil fertility (Diacono and Montemurro, 2010; Martínez-Blanco et al., 2013a,b), defined as the total soil content of plant macro and micronutrients (Havlin, 2013), compared to soils without compost application. Nevertheless, the magnitude of increase depends on compost properties (i.e. maturity, nutrient content, particle size), rate, number of applications, and time of soil sampling (Duong et al., 2013; Liu et al., 2019). Solaiman et al. (2019) studied the short-term change in total soil nutrient content after the application of six different types of compost to a potting soil compared to an unamended control. It was observed that N contents increased between 0% and 40%, total soil P increased between 7% and 13%, total K increased between 35% and 297%, S increased between 7% and 186% and Zn increased between 12.5% and 290% depending on the compost. Remarkably, a study in Mediterranean grasslands by Ryals et al. (2014) showed that the effects of a single compost application on soil fertility can be long lasting and persist in the soil for three years, due to the recycling of plant nutrients and stabilization of the added nutrients.

In cropping systems, due to the generally higher plant nutrient uptake rates and low nutrient recycling rates (i.e. high nutrient exports in harvested products), compost needs to be reapplied regularly to maintain nutrient fertility. After 2 years of compost application to a sugar beet-wheat rotation, Crecchio et al. (2001) reported increases of 2.1% and 7.8% in total soil N, compared to an unamended control, after application of municipal solid waste compost at 120 kg ha^{-1} and 240 kg ha^{-1} respectively (Table 1). Hartl et al. (2003) reported that soil K increased on average by 26% as compared to a control, after 5 years of compost additions. Application of four different types of compost at a rate of 40 Mg ha^{-1} to a sandy soil resulted in increases in soil Olsen-P between 25% and 525% compared to an unamended control (Castán et al., 2016). Eghball and Power (1999) reported increases in soil P of 57%-159% with two different rates of composted manure applied over 4 years as compared to an unamended control.

Long-term (>10 years) application of composted organic waste leads to the buildup of soil organic matter and plant macro- and micronutrients in soil (Table 1) (Chen et al., 2018) to levels sometimes higher than the use of synthetic fertilizers (Lehtinen et al., 2017). It should be highlighted that compost is typically applied at a rate to satisfy the crop N demands which results in overapplication of P. High concentrations of P can be concerning due to possible runoff and leaching and subsequent contamination of the aquifers (Antille et al., 2014; Maltais-Landry et al., 2015). Thus, compost application rates and soil levels of available P need to be closely monitored to avoid losses and subsequent environmental damage through eutrophication of aquatic ecosystems (Chen et al., 2018; Maltais-Landry et al., 2015). Similarly, certain types of compost such as those produced from municipal solid waste can increase micronutrient content to a point where it can cause environmental pollution; in these cases, application rates need to be controlled to avoid environmental pollution (Fagnano et al., 2011).

3.2 Mineralization rates

Most of the nutrients supplied with compost are in organic form and therefore not immediately plant available. Thus, besides the total concentration of nutrients provided with compost, the rates at which these nutrients locked up in organic compounds become available for plants once in the soil through mineralization, is one of the most important components of soil fertility.

Table 1 Examples of changes in common soil health indicators reported in the literature with short- (<1 year), medium- (1–10 year), and long- term (>10 years) compost application, relative to non-amended soils

Soil function	Goal	Indicator	Short-term	Medium-term	Long-term	References
Nutrient cycling	Supply	Available soil N	0% to 40%	2.1% to 7.8%	NA	Crecchio et al. (2001), Eghball and Power (1999), Hartl et al. (2003), Solaiman et al. (2019)
		Soil P	7% to 13%	25% to 525%	NA	
		Soil K	35% to 297%	26%	NA	
		Soil S	7% to 186%	NA	NA	
		Soil Zn	12.5% to 290%	NA	NA	
		CEC	NA	NA	NA	
		pH	NA	NA	NA	
	Recycling and retention	N mineralization	5% to 30%	40% to 50%	50% to 60%	Boldrin et al. (2009), Hartz et al. (2000), Lazicki et al. (2020), Martínez-Blanco et al. (2013a)
		P mineralization	2% to 50%	90% to 100%	90% to 100%	
		K mineralization	75% to 80%	20% to 100%	NA	
Soil hydraulic properties	Water retention/ infiltration	Aggregate stability	18% to 41%	0% to 63%	29% to 238%	Celik et al. (2010), Daynes et al. (2013), Leroy et al. (2008) Mangalassery et al. (2019), Martínez-Blanco et al (2013a), Tejada et al. (2008), Tejada and Gonzalez, (2006a), Xin et al. (2016)
	Water infiltration	Infiltration rate	0% to 300%	24% to 396%	339%	Kranz et al. (2020), Leelamanie and Manawardana (2019), Logsdon et al. (2017), Martínez-Blanco et al. (2013a), Mohammadshirazi et al. (2017), Rivers et al. (2021), Weindorf et al. (2006), Xin et al. (2016)
	Compaction	Bulk density	−2.5% to −55%	0.7% to −40%	−20.6% to −40%	Aggelides and Londra (2000), Courtney and Mullen (2008), Diana et al. (2008), Kranz et al. (2020), Martínez-Blanco et al. (2013a), Mohammadshirazi et al. (2017) Mylavarapu and Zinati (2009), Sax et al. (2017) Stamatiadis et al. (1999)

(Continued)

Table 1 (*Continued*)

Soil function	Goal	Indicator	Short-term	Medium-term	Long-term	References
Crop productivity	Water retention	Water holding capacity	0% to 50%	0% to 2.6%	−0.4% to 1.7%	Arthur et al. (2011), Eden et al. (2017), Leroy et al. (2008), Mamo et al. (2000), Martínez-Blanco et al. (2013a), Miller et al. (2015)
	High crop yield	Crop yield	−48% to 106%	0% to 71%	10% to 27%	Arthur et al. (2012), Chen et al. (2018), Choi et al. (2001), Erhart et al. (2005), Giannakis et al. (2014), Hartl et al. (2003), Iglesias-Jimenez and Alvarez (1993), Mamo et al. (1999), Tejada and Gonzalez, (2006a)
	High crop quality	Vitamin content	24%	NA	NA	Chan et al. (2010), Coria-Cayupán et al. (2009), Martínez-Blanco et al. (2011), Morlat and Symoneaux (2008), Mugnai et al. (2012), Ramzani et al. (2017), Rubio et al. (2013), Saha et al. (2007), Wang and Lin (2003)
		Phytochemicals (phenols, sinapic acid, anthocyanins)	−33% to 76%	NA	−2%	
		Micronutrient content (Zn, Fe)	+12% to 26%	NA	NA	
	Crop health	Disease incidence	−83% to 53%	NA	NA	Avilés and Borrero (2017), Kanaan et al. (2018), Tubeileh and Stephenson (2020)
Soil biodiversity	High microbial abundance	Microbial biomass	0% to 106%	10% to 97%	10% to 400%	Bastida et al. (2007), (2008), Franco-Otero et al. (2012), García-Gil et al. (2000), Lazcano et al. (2013), Li et al. (2015), Nair and Ngouajio (2012), Pezzolla et al. (2013), Ros et al. (2006), Zhen et al. (2014)
	High diversity	Microbial diversity	18% to 34%	NA	3% to 4%	Bastida et al. (2008), Li et al. (2015), Martínez-Blanco et al. (2013a), Zhen et al. (2014)
	High activity	Soil basal respiration	133% to 500%	29%	0% to 43%	García-Gil et al. (2000), Jindo et al. (2016), Li et al. (2015), Pérez-Piqueres et al. (2006), Ros et al. (2003), Tian et al. (2015)
		Enzyme activity	0% to 500%	−18% to 730%	−29% to 300%	

Mineralization of organic compounds is carried out by soil microorganisms and strongly dependent on environmental conditions that regulate microbial activity (i.e. soil temperature, moisture, and pH) (Carvalhais et al., 2014; Frey et al., 2013). These environmental conditions depend on large-scale factors such as the predominant climate and seasonality, but also on small-scale factors such as soil texture. It has been recently suggested that fine-textured soils support higher mineralization rates than coarse-textured soils through mediation of soil water content and its distribution in the soil matrix (Li et al., 2020).

Furthermore, mineralization of organic compounds is strongly dependent on the quality of the organic matter being decomposed. The relative amount and recalcitrance of the organic compounds (i.e. molecular complexity and strength of the chemical bonds) as well as stoichiometric relationships between C and plant nutrients (mostly N and P) can substantially affect the amount of nutrients being released from compost. High recalcitrance and C to N ratio (between 20 and 30) result in the slow release of nutrients and will even decrease the amount of plant-available nutrients in the soil if the C:N ratio of the added compost is above 30 (Lazicki et al., 2020). Immature composts can have large C:N ratios and high amount of labile C compounds that results in a net immobilization of nutrients such as P and N rather than a net increase (Lazicki et al., 2020). The C:N ratio is also strongly associated with the predominant feedstock used in the composting process; typically high C:N composts, such as those produced from yard trimmings, release up to 5% of the N within the first three months after application whereas low C:N composts produced from animal manures can release up to 15-30% of the applied N over the same amount of time (Hartz et al., 2000; Lazicki et al., 2020).

Besides the chemical properties, it has also been suggested that the physical properties of compost and precisely the particle size will influence the total amount of nutrients being released over a certain time. In particular, it has been observed that finer particle sizes typically result in higher mineralization rates than larger particle sizes (Duong et al., 2013). Higher mineralization rates with smaller particle sizes could be attributed to the larger surface area available for microbial attack, lower C:N ratio, and an increase in ash and nutrient content (Haynes et al., 2015).

Finally, nutrient mineralization from compost also depends on the timeframe considered; it has been observed that 5%-22% of the N in compost is mineralized the first year, and 40-50% in 3-5 years, and 50-60% beyond 5 years (Table 1) (Boldrin et al., 2009; Martínez-Blanco et al., 2013a). For P, between 2 and 50% can be mineralized within 1 year and up to 100% within 5 years (Martínez-Blanco et al., 2013a). The existing literature showing release rates of K from compost is scarce; some studies have suggested that release rates reach up to 75-80% in the first year and 20-100% in three to five years of application.

In summary, understanding compost mineralization is essential to predict nutrient availability in soils and synchronize nutrient release with crop nutrient uptake. For this, the quality of the compost amendment, the environmental conditions of temperature and moisture, and the time frame of interest need to be considered to match plant demands and avoid losses of nutrients.

3.3 Soil cation exchange capacity

Cation exchange capacity (CEC) is an important component of soil fertility which defines the capacity of the soil to supply and retain nutrients. CEC depends on the number of negative charges in a specific soil that would be able to adsorb cations, including many essential plant nutrients. These charges are mostly associated with clay minerals and soil organic matter. Thus, increases in soil organic matter content with compost application are associated with higher CEC as compared to unamended soils (Miller et al., 2015; Weber et al., 2007). Previous research shows variable effects of compost on CEC, which may depend on the rate of compost application, time since application and characteristics of the compost such as compost maturity and feedstock (Harada and Inoko, 1980; Shiralipour et al., 1992). For example, increasing application rates from 20 to 80 Mg ha^{-1} y^{-1} can increase soil CEC of a soil between 7% and 18% respectively compared to unamended soil (Table 1) (Miller et al., 2015). However higher increases in soil CEC can be expected with aged or mature compost compared to fresh compost (Liu et al., 2019).

3.4 Soil pH

Soil pH drives the solubility of elements including plant nutrients, being a very reliable indicator of plant nutrient availability in soils. Plant macronutrients are mostly soluble at pH 6.5-7.5, becoming unavailable at higher or lower pH through complexation with other elements in the soil solution. For instance, phosphate (HPO_4^{-2} or $H_2PO_4^{-}$), the plant available form of P, co-precipitates with calcium (Ca^{2+}) at pH higher than 7 and with aluminum (Al^{3+}) and iron (Fe^{3+}) when pH is below 5 (Brady and Weil, 2016). Therefore, slightly acidic conditions are desired to avoid nutrient deficiencies. The literature shows inconsistent effects of compost on soil pH. For example, depending on compost stability, short-term decreases in soil pH can be observed due to organic acid production during the decomposition of labile organic molecules and oxidation of N and S compounds (Franco-Otero et al., 2012). However, other studies reported slight increases in soil pH shortly after compost addition (Tian et al., 2015) and in fact, compost has been used successfully to buffer soil acidity in degraded, soils in tropical (Latifah et al., 2018; Ouédraogo, 2001) and temperate regions of the world (Domínguez et al., 2019). Reports on the long-term effects of compost application on soil pH are also inconsistent showing increases (Butler and Muir, 2006; García-Gil et al., 2004), decreases (Bastida et al., 2008; Meng et al., 2005), or no differences after 13 years of continuous application of compost (Calleja-Cervantes et al., 2015). Besides compost stability, other factors that seem to be important in driving the effects of compost on soil pH are the pH of the compost and application rates (Diacono and Montemurro, 2010).

4 Effects of compost on soil hydraulic properties

4.1 Soil aggregate stability

Soil structure regulates the balance between macro and micropores and therefore has a strong influence in soil hydraulic properties, allowing for the balancing of a soil's water holding capacity and infiltration, reducing soil erosion. Furthermore, a good soil structure facilitates plant root growth, creates habitat for soil biota (Erktan et al., 2020), and leads to C sequestration (Six et al., 2000). The formation of aggregates through the binding of soil particles by organic compounds is one of the major processes contributing to the creation of structure in the topsoil (Bronick and Lal, 2005). Compost can increase the formation of water-stable aggregates directly by supplying organic compounds (i.e. polysaccharides, proteins, lipids) that will increase particle inter-cohesion and aggregate hydrophobicity or indirectly through the increase in microbial biomass and the subsequent generation of microbial residues, known to be excellent binding agents. Several studies show that compost application can increase aggregate stability which depends on the quality of the compost, the number of consecutive applications, and length of the experiment (Table 1) (Diacono and Montemurro, 2010). For instance, application of compost with high content of easily decomposable organic compounds such as glucose and other monosaccharides leads to intense and transient increases in aggregate stability, whereas composts with high contents of recalcitrant compounds such as lignin lead to lower but longer lasting effects (Abiven et al., 2009). In an incubation experiment, Temiz et al (2021) observed that aggregate stability was increased up to 18% compared to a control only 60 days after compost application (Table 1). Nevertheless, this depended on the compost feedstock and application rates, with compost produced from animal manures and higher compost application rate leading to the largest increases. In a review of the literature published between 1990 and 2012, Martinez-Blanco et al. (2013a,b) pointed out higher increases in aggregate stability (29–41%) in the short-term (<1 year after compost application). The short-term formation of stable aggregates after compost application is further accelerated by the presence of plant roots and arbuscular mycorrhizal fungi, which suggests potential synergies and highlights the importance of soil agroecological management (Celik et al., 2010; Daynes et al., 2013). In the medium term, previous studies showed either no changes in aggregate stability (Tejada and Gonzalez, 2006a) or significant increases of 10.5%, 37.8% or 63% compared to a control (Leroy et al., 2008; Mangalassery et al., 2019; Tejada et al., 2008). In the long term, it was observed that repeated compost application for 13 and 23 years increased the proportion of water stable aggregates by 29% and 238%, respectively (Celik et al., 2010; Xin et al., 2016).

4.2 Water infiltration rate

A good soil structure and porosity are essential to reduce ponding and facilitate the movement of water into the soil profile (i.e. infiltration). By improving aggregation and infiltration rates, compost application facilitates the recharge of aquifers and reduces soil erosion rates (Martínez-Blanco et al., 2013b). The published literature shows variable effects of composts on soil infiltration rates in the short term (<1 year after compost application), with either no observed effects (Leelamanie and Manawardana, 2019; Weindorf et al., 2006) or increases of 300% compared to a control soil (Table 1) (Rivers et al., 2021). In the mid-term, between 1 and 10 years after compost application, studies report increases between 24% and 396% compared to a control (Kranz et al., 2020; Logsdon et al., 2017; Mohammadshirazi et al., 2017). In the long term, Xin et al. (2016) reported increases of 339% in infiltration rates compared after 23 years of compost application compared to a control soil without compost.

4.3 Compaction

Soil compaction can be assessed through the measurement of soil bulk density or penetration resistance in the field. By promoting aggregate formation, and increasing soil organic matter content, compost can increase soil porosity which is associated with lower bulk density and penetration resistance (Diacono and Montemurro, 2010; Ruehlmann and Körschens, 2009). In the short term, i.e. <1 year, the application of different types of compost decreased soil bulk density between 2.5% and 55% as compared to soil without compost (Table 1) (Aggelides and Londra, 2000; Courtney and Mullen, 2008; Diana et al., 2008; Kranz et al., 2020; Mohammadshirazi et al., 2017; Mylavarapu and Zinati, 2009; Stamatiadis et al., 1999), while in the mid-term (1–10 years after compost application), the decrease may be as much as 40%, with a minimum decrease of 0.7% (Kranz et al., 2020; Martínez-Blanco et al., 2013b). Long-term (> 10 years) compost application can decrease soil bulk density between 20.6% and 40% (Kranz et al., 2020; Martínez-Blanco et al., 2013b; Sax et al., 2017). Compost application seems to be more effective in the short term (Weber et al., 2007) and with increasing application rates (Hemmat et al., 2010), as larger doses produce larger decreases in bulk density.

4.4 Water holding capacity

Increased drought severity due to climate change is increasing the interest in improving water use efficiency in crop production. Besides the use of water-efficient crops and irrigation systems, improving soil water holding capacity is considered as a key strategy to adapt to current and future water shortages. Soil water holding capacity (WHC) is linked to porosity and therefore aggregate formation and soil structure are critical in soil water retention. Given the important role of soil organic matter in driving aggregate formation and soil porosity, it has been observed that increasing soil organic matter leads to the increase in soil WHC (Rawls et al., 2003). In a meta-analysis of the published literature, Minasny and McBratney (2018) observed that a 1% increase in soil organic C was associated with significant increases in soil plant-available water-holding capacity, yet these increases were highly dependent on soil texture. Consequently, the effects of compost on soil WHC observed in the literature are also highly variable (Table 1). Martínez-Blanco et al. (2013a,b) showed that, in the short term, compost application resulted in 0–50% increases in soil WHC. In the medium term, consecutive applications of 90 tn ha^{-1} yr^{-1} compost to corn over 3 years have shown to produce no changes in soil WHC (Mamo et al., 2000), whereas Ramos (2017) observed that compost application for 3 years resulted in increases between 1.2% and 2.6% of soil WHC in a Mediterranean vineyard. In the long term, subsequent applications of compost for more than 10 years have shown to either decrease soil WHC by 0.4% (Leroy et al., 2008) or increase soil WHC by between 0.7% and 1.7% (Arthur et al., 2011; Eden et al., 2017; Miller et al., 2015). This large variability in the observed results is due not only to the soil texture but also to the application rate and particle size of the compost applied. Higher rates and finer particle sizes result in larger increases in soil WHC (Głąb et al., 2020; Şeker and Manirakiza, 2020).

5 Effect of compost on crop productivity

5.1 Crop yield

Crop yield, the mass of marketable crop produced per unit of surface area, is often the single most important variable for agricultural producers. The effects of compost application on crop yield depend largely on the crop, timeframe considered, compost properties, compost application rates, baseline soil fertility, characteristics of the control treatment, and environmental conditions (moisture and temperature). In a meta-analysis of 53 studies, Wortman et al. (2017) reported that larger yield increases are observed in leafy greens and annual crops in the first season of compost application compared to root/tuber/bulb crops, especially when the compost had relatively low C:N ratio, and was applied to soils with initial low fertility. This is due to the fact that leafy greens and herbs are often more responsive to increasing fertility (either organic or synthetic) than fruiting, grain, or root crops (Wortman, 2015).

In one single season, compost application has shown to increase the yield of ryegrass between 30% and 106% compared to a control, depending on the application rate (Table 1) (Iglesias-Jimenez and Alvarez, 1993). Nevertheless yield decreases between 35% and 48% have been observed in crops such as corn, tomatoes, and lettuce, due to the immobilization of N when compost with high C:N ratio is used (Choi et al., 2001; Giannakis et al., 2014). In the medium term, subsequent applications of compost for 3 years increased the yield of a rice crop between 5% and 7% (Tejada and Gonzalez, 2006b) and corn by 71% compared to a control (Mamo et al., 1999). Nevertheless, Hartl et al., (2003) saw no effects on the yield of a rye crop after 5 years of compost application.

Long-term studies also show variable effects of compost on crop yields. For example, Arthur et al. (2012) observed an increase in tomato yields of 27% after 10 years of subsequent compost additions (7.5 ton ha^{-1} yr^{-1}) as compared to a control. Erhart et al. (2005) observed 10% yield increases after 10 years of compost amendment to cereals and potatoes. In a recent meta-analysis, Chen et al. (2018) pointed out that the effects of long-term application of compost for crop yield depends on compost-, soil-, and climate-dependent factors. Soils with low initial fertility, near neutral pH, low soil organic matter, finer texture, and in tropical climates show the highest increases in crop yield after long-term application of organic fertilizers (Chen et al., 2018). Consistently, several studies show that compost use is a great alternative in low-input tropical cropping systems such as sorghum in West Africa, or maize, wheat, potato, and faba bean smallholder farms in Ethiopia, where access to synthetic fertilizers is limited (Bedada et al., 2014; Ouédraogo, 2001). Finally, it should be highlighted that the effects of compost on crop yield can be smaller than those of equivalent amounts of synthetic N, P, and K inputs due to the slower mineralization and nutrient release in compost (Martínez-Blanco et al., 2013a). Similar yields can be achieved when compost is supplemented with inorganic fertilizers. Furthermore, some studies have reported larger increases in crop yield and nutritional quality with compost as compared with equal amounts of synthetic fertilizers, which could be attributed to the supply of secondary macronutrients or micronutrients, as well as improvement of soil biological and physical properties (Celestina et al., 2019; Lazcano et al., 2013).

5.2 Crop quality

Scientific evidence shows that the importance of a healthy diet for health and wellbeing has increased consumer and farmer interest in the nutritional quality of crops (Willett et al., 2019). High concentrations of macro- and micronutrients, sugars, phytochemicals, and secondary metabolites in agricultural products are generally desired, although the parameters that define crop quality are crop dependent (Brandt et al., 2011). Because of this, it is extremely difficult to summarize and compare current literature studying the effects of compost on crop quality. Generally, it has been observed in large-scale studies that continued application of high-analysis simple synthetic fertilizers and export of harvest products can lead to the gradual depletion of micronutrients such as Zn and Fe in soils (Chen et al., 2017). Even if this does not negatively affect crop growth and yield, it can decrease the micronutrient concentration in the harvested parts of the plant and therefore the crop nutritional value (Chen et al., 2017). This decrease in crop nutritional value, called 'hidden hunger,' is further exacerbated by increasing crop yields due to a 'dilution effect' (Liu et al., 2014). On the other hand, higher crop nutritional quality has been observed in several organic systems that use organic fertilizers, an effect which is attributed to the supply of both macro and micronutrients, and the lower crop yields that avoid the abovementioned 'dilution effect' (Helfenstein et al., 2016; Lairon, 2011). Several studies have evaluated the effects of compost on crop nutritional quality (Table 1). In the short term (< 1 year after compost application), Ramzani et al. (2017) showed that application of small amounts of acidified compost to a saline soil could increase Zn and Fe concentrations in grain by 26% and 12.5% respectively, compared to a control. Furthermore, compost increased grain protein content by 3%, vitamins by 24%, although polyphenol concentration decreased by 33% compared to the control. In rice, Saha et al. (2007) observed that compost application increased grain Zn and Fe contents by 14 and 12% respectively, compared to inorganic fertilizers. In strawberries, Wang and Lin (2003) reported significant increases in antioxidant activity and flavonoid contents in plants grown in pots with 50% soil and 50% compost as compared to plants grown in pots with 100% soil. In lettuce, Coria-Cayupán et al. (2009) reported higher contents of chlorophylls and carotenoids, but lower contents or non-significant differences for phenol compounds and antiradical activity. In cauliflower, Martínez-Blanco et al. (2011) observed up to 76% and 24% higher content of sinapic acids and phenols, respectively, in plants grown with compost compared to inorganic fertilizers. In winegrapes, a single application of compost produced from exhausted grape marc, citrus juice waste, and cattle manure resulted in 9% increase in tartaric acid and 16% increase in Fe in fruit, while decreasing total anthocyanins, polyphenols, and Zn by 1%, 9%, and 7% respectively as compared to an unamended control (Rubio et al., 2013).

Consecutive annual applications of compost to perennial crops have also shown inconsistent effects on crop quality. For example, application of green waste compost to Chardonnay winegrapes for 8 years resulted in no change in soluble solids (measured as °Brix), titratable acidity, malic and tartaric acid concentrations compared to inorganic fertilizers, parameters than are known to influence grape fermentation and wine quality (Mugnai et al., 2012). Chan et al. (2010) observed that application of compost to winegrapes for three consecutive years resulted in only slight increases (0.4%) in grape pH. In the long term, application of compost to winegrapes for 28 consecutive years reduced the content of total soluble solids (°Brix) by 4%, anthocyanins by 2%, and tannin content by 9%, and increased pH by 1.7% compared to an unamended control (Morlat and Symoneaux, 2008). Thus, there is evidence that compost use can change crop quality, although general conclusions cannot be made, given the large variability in the observed results and diversity of quality criteria used in each crop. Furthermore, there is a shortage of studies evaluating the long-term effects of compost on crop quality.

5.3 Crop health

Pests and diseases can greatly decrease the yield and quality of crops and are therefore one of the main concerns of agricultural producers. Conventional agriculture relies in heavy applications of pesticides which in many cases have shown to be harmful for soil, environmental, and human health. Furthermore, repeated applications of these pesticides lead to the generation of resistance in the targeted pathogens and a decrease in the efficiency of the fumigants over time (Lucas et al., 2015). The capacity of compost to suppress or reduce the incidence of plant diseases has been demonstrated in numerous studies and is extensively reviewed in (Bonanomi et al., 2007; De Corato, 2020; Martínez-Blanco et al., 2013a; Noble and Coventry, 2005). Compost can support plant health indirectly, by supplying the required nutrients and providing favorable conditions for plant growth. In addition, compost can directly suppress the presence of several soil-borne plant pathogens and increase the resistance of the plants by inducing systemic resistance (Noble and Coventry, 2005). Direct suppression is known to be well correlated to changes in soil microbial community structure and function with compost application and with the presence of microbes that directly compete with the soil-borne pathogens or that produce antibiotic substances (Bonanomi et al., 2010; De Corato, 2020; Mehta et al., 2014; Weller et al., 2002). Given the large differences in microbial and physicochemical properties of different types of compost, it is expected that suppressiveness is also highly variable (Table 1). For instance, in a pot experiment, Kanaan et al. (2018) reported a decrease of 83% in disease severity of *Verticillium dahliae* in eggplants grown in compost produced from tomato waste, but increase of 53% when plants were grown with compost produced from olive waste, as compared to a non-amended peat potting media. In a different pot experiment comparing four types of olive waste compost, Avilés and Borrero (2017) saw both increases and decreases in *V. dahliae* disease severity in cotton seedlings, depending on the type of compost. Finally, compost suppressiveness is often pathogen specific. For instance, compost has shown higher suppression rates in pathogens with a limited saprophytic capacity such as *Thielaviopsis basicola* and *V. dahliae* but low suppression rates or even increases in the abundance of the soil-borne fungi *Phytophthora* spp., *Rhizoctonia solani*, and *Pythium* spp. (Bonanomi et al., 2010). In a field study with bell peppers, Tubeileh and Stephenson (2020) observed that plant-based composts reduced the incidence of *V. dahliae* by 60%, while compost produced from dairy manure was not suppressive. Furthermore, disease suppression was transient and disappeared 2 weeks after compost application. In summary, while the use of compost to suppress soil-borne diseases seems like a promising strategy, much research is still needed to determine the specific mechanisms of suppression, and general management recommendations cannot be made.

6 Effects of compost on soil biodiversity

Soil biota encompasses a wide diversity of organisms, represented across all life kingdoms, and ranging in size from microorganisms such as bacteria and fungi to megafauna such as moles or ground squirrels. The abundance of soil biota can be extremely high, particularly in the case of microorganisms, with billions of microbial cells per gram of soil (Bardgett and van der Putten, 2014). By supporting the major processes of decomposition, nutrient cycling, and soil formation, soil biota is crucial for the maintenance of soil health and crop productivity (Bardgett and van der Putten, 2014; de Vries et al., 2013; Lehman et al., 2015). Thus, a mechanistic understanding of the effects of compost on soil health necessarily involves a thorough study of how it affects soil biota. For this, the development of reliable, robust, and informative soil biological indicators is of high importance for the scientific community. Soil biological communities can be studied from the perspective of *'who is there'* or community structure, and *'what are they doing'* or community function. Community structure is described as both the number of different species that form a specific community, and their biomass. Measurements of function of soil communities include microbial respiration (Degens and Harris, 1997), stable isotope probing (Neufeld et al., 2007), reverse transcription of functional genes coding for specific enzymes (Sharma et al., 2005), or potential activity of hydrolytic enzymes (Fierer et al., 2021; Nannipieri et al., 1980; Tabatabai, 1994).

6.1 Microbial biomass

Microbial biomass constitutes only a small percentage (1–5%) of the total soil organic matter, yet it is extremely important for soil functioning supporting not only nutrient cycling, but also acting as a reservoir of nutrients and a vehicle for C sequestration (Zhang et al., 2017; Zhu et al., 2020). Soil microbial biomass can be estimated as the sum of the concentration of phospholipid fatty acids (PLFAs) for bacteria and fungi, the concentration of the biomarker ergosterol for fungi, through substrate-induced respiration, adenosine triphosphate (ATP) concentration, or as total microbial biomass – C (MBC), N (MBN), measured through fumigation-extraction (Frostegård et al., 1993; Horwath and Paul, 1994; Vance et al., 1987; Young, 1995). Even though it has been observed that a compost with poor quality (i.e. with heavy metals or high salinity) can decrease microbial biomass (de Araújo et al., 2010), typically, the organic compounds found in compost are a source of energy and nutrients for soil microorganisms. Therefore, compost application can boost microbial growth in the short term (Table 1). In a 45-day incubation study, Pezzolla et al. (2013) observed that application of municipal solid waste compost increased soil MBC up to 55% relative to an unamended control. In a pot experiment, Zhen et al. (2014) observed that manure compost increased soil microbial biomass by 58% in just one season. In a field experiment, Bastida et al. (2008) showed that composted sewage sludge could increase MBC by 106% in only one season. Nevertheless, Franco-Otero et al. (2012) did not see increases in MBC after application of municipal solid waste or sewage sludge compost in a semi-arid Mediterranean soil. Similarly, Lazcano et al. (2013) reported no increases in total PLFAs after adding vermicompost to a corn crop. These differences in the short-term effects of compost application observed in the literature have been attributed to the relative proportion of labile and recalcitrant C compounds in the compost (Calbrix et al., 2007; Lazcano et al., 2013).

Studies with repeated applications of compost over several years show more consistent positive effects on microbial biomass (Table 1). For example, two years of compost application to a Mediterranean soil resulted in a 97% increase in MBC compared to an unamended control (Bastida et al., 2008). Nair and Ngouajio (2012) reported a 50% increase in MBC after 3 years of dairy compost application at 25 t ha^{-1}. Similarly, García-Gil et al. (2000) observed that consecutive annual applications of 20 t ha^{-1} or 80 t ha^{-1} of municipal solid waste compost for 9 years increased MBC by 10% and 46%, respectively. In the long term, it was observed that 17 years of annual applications of compost to degraded semi-arid soil resulted in 84–400% increases in MBC, depending on the application rate (65 t ha^{-1} and 260 t ha^{-1} respectively) (Bastida et al., 2007). Ros et al. (2006) observed increases of up to 10% in MBC after 12 years of green waste compost application at 17 t ha^{-1} yr^{-1}. Li et al. (2015) reported increases of 83% to 138% in MBC after 25 years of continuous application of composted cattle manure at 375 kg N h^{-1}yr^{-1} and 750 kg N h^{-1}yr^{-1}, compared to an unamended control in a corn–wheat rotation.

6.2 Microbial diversity

Microbial diversity is an important component of soil health since it is responsible for the provision of ecosystem service and resilience to external disturbances (Wagg et al., 2014). Changes in microbial diversity induced with soil management have been associated with the overall decline in the function of the microbial community (Tsiafouli et al., 2015). Soil microbial diversity can be assessed with different molecular techniques. Fingerprinting DNA-based techniques such as terminal restriction fragment length polymorphism (T-RFLP) and denaturing gradient electrophoresis (DGGE) were commonly used in the 1990-2000s. Recently, high-throughput sequencing of gene markers, such as 16S rRNA (for prokaryotes) and ITS (for fungi), is commonly used in assessments of soil health (Fierer et al., 2021; Lindahl et al., 2013; Ward et al., 1990). The diversity of soil microbial communities is frequently calculated through indices that take into account species richness and the relative abundance of each species (i.e. Shannon diversity index, Simpson, ACE, Chao 1) (Kim et al., 2017).

Compost has short-, medium-, and long-term effects on soil microbial diversity (Table 1) (Martínez-Blanco et al., 2013b). Using DGGE, Zhen et al. (2014) reported that, in one single season, compost addition could increase bacterial and fungal diversity by 34% and 18% respectively compared to an unamended control. Studies show that, generally, in the short term, gram-negative and gram-positive bacteria are stimulated by the addition of labile C compounds with compost (Lazcano et al., 2013; Willekens et al., 2014) whereas fungi are not so strongly affected (Lazcano et al., 2013). This selection of soil microorganisms results in differences in microbial community structure between compost amended and unamended soils observed within only one season (Farrell et al., 2010; Pérez-Piqueres et al., 2006; Watts et al., 2010). The changes exerted on the composition of the microbial community by a single application of compost can be long-lasting and visible 24 years after as shown by Torres et al. (2015).

Repeated applications over more than 1 year can also shift the composition of the microbial community. Two years after applying sewage sludge compost to degraded soil, Bastida et al. (2008) reported significant changes in the structure of the soil microbial community as assessed by PLFAs, a change that was associated with higher microbial biomass activity and soil C. Nevertheless, other studies show no differences in bacterial community structure between compost-amended soils and controls even after 5 years of subsequent compost applications (Cherif et al., 2009). In the long term, 25 years of continued applications of composted cattle manure at 375 kg N ha^{-1}yr^{-1} and 750 kg N ha^{-1}yr^{-1} to a corn-wheat rotation increased bacterial diversity by 3% and 4% respectively, compared to an unamended control in a corn-wheat rotation (Li et al., 2015). The contrasting results observed in the effects of compost on soil microbial diversity have been attributed to various factors such as the application rate (higher application rates result in stronger changes), and the proportion of mineralizable carbon and lignin in the applied compost (Saison et al., 2006).

6.3 Microbial activity

Measurements of microbial activity are often preferred to measurements of microbial diversity or community structure since they are directly correlated to soil processes such as mineralization of organic compounds and release of plant-available nutrients (Kibblewhite et al., 2008). Microbial activity can be estimated, among other methods, by measuring soil respiration or potential activity of microbially produced enzymes involved in the catabolism of organic compounds (Degens and Harris, 1997; Nannipieri et al., 1980; Tabatabai, 1994). Both microbial respiration and enzyme activity are sensitive to changes in soil management and have shown to change with compost application (Martínez-Blanco et al., 2013b). High concentration of heavy metals and other pollutants in compost can decrease soil microbial activity; however, if compost quality is good, typically an increase in microbial activity is expected after compost application (Knapp et al., 2010). In addition to estimating overall microbial activity, many studies assess the diversity of metabolic functions supported by the soil microbial community (i.e. functional diversity) using community-level physiological profiles or CLPPs. The use of BIOLOG Ecoplates® allows for the simultaneous analysis of 31 C sources using a tetrazolium redox dye as an indicator of microbial metabolism (Garland and Mills, 1991). This method stems from the early work of Degens and Harris (1997) who developed a method to estimate substrate-induced respiration using an array of C sources of different complexity.

In a short-term incubation experiment with two soils, Pérez-Piqueres et al. (2006) observed large increases (between 133% and 500%) in soil basal respiration after application of compost as compared to an unamended control (Table 1). Furthermore, this same study reported significant changes in the CLPPs in amended soils as compared to unamended soils only 10 days after the start of the experiment, indicating a rapid shift in the capacity of the bacterial community to degrade C compounds. In a field experiment, Ros et al. (2003) observed that right after addition of 30 kg compost m^{-2} to a degraded soil, microbial respiration increased by 500%, dehydrogenase activity by 300%, protease activity by 200%, but did not change glucosidase activity compared to an unamended control.

In the medium term (1-10 years of compost application), Jindo et al. (2016) reported that 5 years of sheep manure compost application to plum trees at 20 t ha^{-1} yr^{-1} resulted in a 29% increase in microbial respiration but

18% decrease in dehydrogenase enzyme activity, as compared to the control. Application of composted manure for 9 consecutive years increased the activity of dehydrogenase and catalase by 730% and 200% (García-Gil et al., 2000). Tian et al. (2015) reported that microbial activity measured as hydrolysis of fluorescein diacetate increased by 250% after 3 years of application of 300 kg ha^{-1} yr^{-1} of composted cattle manure.

In a long-term experiment, Ros et al. (2006) reported significant increases in microbial activity with annual application of compost for 12 years, although this depended on the type of compost. Sewage sludge compost increased basal respiration by 43% whereas municipal solid waste and green waste compost did not have any effects. Li et al. (2015) reported that annual application of composted manure for 25 years increased urease activity by 300%, alkaline phosphatase by 175%, ⍰-Glucosidase by 65%, ⍰-Glucosidase by 175%, ⍰-Xylosidase by 225%, ⍰-Cellobiosidase by 280%, and L-Leucine Aminopeptidase by 25%, compared to non-amended soil. This increased enzyme activity was attributed to higher soil organic C and microbial biomass in the plots amended with composted manure. Nonetheless, two other enzymes, peroxidase and phenol oxidase, were decreased by 12.5% and 29% compared to the non-amended soil, a decrease that could be attributed to higher N inputs and lower fungal abundance compared to the non-amended control (Li et al., 2015).

7 Effects of compost on environmental quality

7.1 Carbon sequestration

The global biophysical potential to decrease greenhouse gas emissions by building soil C was estimated at an upper limit of 4–5 Gt CO_2 per year, if soil conservation practices were extensively adopted worldwide (Paustian et al., 2019). Compost application to agricultural lands is among the soil conservation practices well known to sequester C. In fact, several studies that synthesized data on the impact of conservation practices showed greater soil carbon sequestration rates associated with the addition of organic amendments, including compost, compared to other practices such as cover cropping or reduced tillage intensity (Payen et al., 2021; Tiefenbacher et al., 2021). Compost application directly impacts soil C by adding organic matter to the system, as approximately half of the organic matter in compost is C. Compost can also indirectly impact soil C inputs by stimulating plant growth. Not all carbon added as compost to the soil is sequestered. Upon application, compost undergoes decomposition, during which part of the C is respired and lost as CO_2. A culmination of research findings in recent decades supports the concept that soil organic matter is a continuum of progressively decomposing organic compounds, during which C can be stabilized in pools with relatively long turnover times, for example inside aggregates or adsorbed onto soil mineral surfaces (Lehmann and Kleber, 2015). Thus, the amount of C sequestered following compost application depends on the amount of C that is respired versus assimilated into biomass, and the amount of C that can be stabilized in pools with slower turnover times. These processes, in turn, are affected by soil physical, chemical, and biological properties as well as the quality and quantity of the organic matter inputs. Therefore, it is not surprising to see a wide range in soil C sequestration rates following compost application (Table 1). For example, a global meta-analysis of 92 studies in which organic amendments had been added to arid, semiarid, or Mediterranean rangelands showed a net increase in soil organic carbon following compost application (Gravuer et al., 2019). However, the magnitude of gains in soil organic carbon was affected by complex interactions between climate, the compost nitrogen concentration, and the number of years following compost application. Given the limited supply of compost relative to the acreage of agricultural lands, it is pertinent to carefully assess under which conditions compost application can lead to the greatest soil organic carbon gains.

7.2 Nitrogen and phosphorus losses

Nitrogen and phosphorus are essential nutrients for plant growth but cause major environmental problems when lost from agricultural fields. Given that compost contains N and P, compost can be used as fertilizer source, but also runs the risk of causing environmental pollution when excess N or P are applied. When compost is used as a fertilizer, it is typically applied at a rate to meet the crop's N requirement. This is particularly prominent in organic cropping systems, where compost can be the dominant nutrient source. For example, several long-term experiments comparing organic and conventional practices have used compost as dominant nutrient source in treatments under organic management (Krause et al., 2020; Tautges and Scow, 2020). In such experiments, unbalanced soil N:P ratios have been observed after long-term compost application (Maltais-Landry et al., 2016). The dominant form of N in compost is organic N, which needs to be mineralized before N becomes plant available. As such, N applied in the form of compost is typically more slowly available to the plant compared to synthetic fertilizer N. In comparison to soils that receive no N-containing amendments at all, compost application can enhance N leaching (Li et al., 1997). However, when compost is used instead of mineral fertilizer N, reduced N leaching losses are typically observed (Siedt et al., 2021). The amount of N leached appears to be positively correlated to the C:N ratio of the compost (Leclerc et al., 1995; Li et al., 1997). This is in line with close correlations observed between the C:N ratio of organic amendments and associated N mineralization rates (Lazicki et al., 2020). To minimize N leaching after compost application, it is recommended to closely match potentially mineralizable compost N to crop N requirements, considering potential legacy effects of previous years' compost inputs.

As opposed to N, compost P can be highly soluble and more easily lost from the soil compared to its inorganic counterpart (Bernal et al., 2017). This is compounded with nutrient imbalances, where P application often exceeds crop demand by matching compost application rates to meet the crop's N requirement (Eghball, 2003). In a column experiment using three soils and seven livestock manure composts, P leaching was positively correlated with the total P concentration of the compost as well as the soil, and negatively correlated with the P sorption capacity of the soil (Kim et al., 2011). In another study, P leaching increased with increased simulated rainfall and losses were greater following application of dairy compost compared to raw dairy manure (McDowell and Sharpley, 2004). In general, less research has been conducted on the effect of compost application on P loss compared to N loss. However, it is recommended that compost application rates are carefully selected to prevent excessive P inputs, especially in soils with high soil P content or low P sorption capacity.

7.3 Greenhouse gas emissions

The effect of compost application on the net greenhouse gas (GHG) budget does not only depend on soil carbon sequestration, but also on emissions of nitrous oxide (N_2O) from soil, as well as potential upstream emissions (CO_2, N_2O, and methane (CH_4)) associated with compost production and distribution. Nitrous oxide is a GHG with a potency 265 times that of CO_2 on a 100-year horizon (IPCC, 2014), predominately emitted from agricultural soils through the microbially mediated processes of nitrification and denitrification (Butterbach-Bahl et al., 2013). The strong correlation between N_2O and N application has led to the development of emission factors to quantify N_2O emissions based on land application of N inputs for national inventory purposes (Solomon et al., 2007). Extensive literature review settled on a default emission factor of 1%, meaning that for each kilogram of N applied, 0.01 kg of N_2O-N is estimated to be emitted (Solomon et al., 2007). Following the release of the 1% default emission factors, various efforts were geared towards developing more context-specific emission factors, including emission factors that differentiate emissions associated with organic versus synthetic fertilizer N inputs. In general, N_2O emissions appear to be lower following application of compost compared to synthetic N fertilizer or unprocessed organic fertilizers such as raw manure (Lazcano et al., 2021b; Walling and Vaneeckhaute, 2020). However, between-study variability is large. For example, a recent literature synthesis showed N_2O

emission factors for compost ranged between 0.06% and 5.60% across studies (Walling and Vaneeckhaute, 2020). Clear patterns to explain variability between studies are yet to emerge, but the variability is likely attributed to variation in soil type, climate, compost quality, and mode of compost application. Compost is often applied as a soil amendment, meant to improve soil health rather than meeting the full crop nutrient requirement. In these situations, compost is often complemented with synthetic fertilizer N inputs, to meet both goals of building soil health and ensuring immediate crop N requirements. The combined use of compost and synthetic fertilizer N and lack of a consistent control treatment across studies further complicate unraveling the effect of compost on N_2O emissions (Graham et al., 2017). The large variability in effects of compost on soil C sequestration as well as N_2O emissions underlines the importance of further research into potential tradeoffs between C sequestration and N_2O emissions at the field scale (Graham et al., 2017; Walling and Vaneeckhaute, 2020). To assess the effect of compost application on GHG emissions beyond the field scale, upstream emissions associated with compost production and distribution need to be accounted for, as well as avoided emissions associated with alternative management of the organic waste feedstock. Moreover, the net effect on GHG emissions will also depend on whether compost is used as substitute for synthetic fertilizer, replaces raw manure inputs, or is simply added to the standard nutrient management plan. Greenhouse gas emissions during compost production vary widely but can be reduced through strategic selection of the composting method (Ba et al., 2020; Walling and Vaneeckhaute, 2020). For example, silo composting, adding sawdust or straw, and good aeration can reduce GHG emissions during the composting process. Given that typical compost application rates range between 4 Mg ha^{-1} and 40 Mg ha^{-1}, GHGs associated with transportation from the composting facility to the field site add up quickly (Tiefenbacher et al., 2021; Walling and Vaneeckhaute, 2020). Therefore, it is recommended that compost is locally sourced.

8 The use of compost to improve soil health in annual crops: a case study with strawberries

Cultivated strawberry (*Fragaria x ananassa* Duchesne ex Rozier) is a fruit crop highly appreciated throughout the world and of major economic importance in countries such as China, USA, Mexico, Egypt, Turkey, and Spain. One of the biggest challenges in strawberry cultivation is the management of soilborne diseases. Since the 1960s, methyl bromide has been almost universally used to control soilborne diseases in conventional strawberry (Chellemi, 2002), but this fumigant has been phased out of production and use as of 2016 due to its negative impacts on public health and the environment. To combat the increased disease pressure in strawberry production, researchers are proposing an integrated approach including breeding for increased disease resistance, development of novel pesticide chemistries, optimization of cultivation practices, and improvement of the health and disease-suppressiveness of the soil (Lloyd, 2016). To date, strawberry cultivars have been identified that are tolerant to soil-borne pathogens like *Verticillium dahliae* (Shaw et al., 2010, 1997) and *Fusarium oxysporum* f. sp. *fragariae*. Additionally, the strawberry industry is looking for cultivars that are tolerant to *Macrophomina phaseolina*, an emerging disease that causes crown rot in strawberries and is seriously threatening the strawberry industry in California (Koike, 2008). The mechanisms of disease tolerance are still unclear, but recent research points out at possible interactions between resistant cultivars and beneficial microorganisms colonizing in the rhizosphere (Lazcano et al., 2021a).

Another major challenge in strawberry production where soils play a critical role is the management of nitrogen (N). California strawberry growers commonly apply N fertilizer in two forms: a pelletized, coated pre-plant fertilizer or control release fertilizer (CRF) and in-season liquid N fertilizer via drip irrigation, a method called fertigation. California strawberry fields typically receive between 168 kg ha^{-1} and 336 kg ha^{-1} fertilizer N over a production season while strawberry crop N uptake has been reported to range from 201 kg ha^{-1} to 246 kg ha^{-1}, but N fertilization rates and timing vary widely among growers. Nitrate pollution of freshwater resources associated with leaching of fertilizer nitrogen from agricultural lands is a major challenge in many areas of the world where strawberries are grown (Carle et al., 2006). The strawberry industry has focused on 'fine-tuning' fertility and water management for more efficient resource use (Bottoms et al., 2013a,b). Several studies reported a high potential for leaching of N fertilizer during the early stages of plant development, when plant uptake is low (Bottoms et al., 2013b). Thus, there is a strong need to find slow-release fertilizers that supply enough nutrients to support crop growth while reducing the risks of N loss.

By improving soil health, compost application presents one potential strategy to address these two major challenges in strawberry production. Long-term management of soils with organic amendments increases soil organic matter, microbial biomass, and activity compared with fertilized or unamended soils (Diacono and Montemurro, 2010). By increasing the size and activity of the soil microbial community, compost could be conducive to the establishment of beneficial plant-microbial interactions, resulting in improved plant health and nutritional quality. In fact, compost has been shown to suppress soilborne disease in several published studies (Noble and Coventry, 2005). Lodha et al. (2002) found that amendment of soil with compost prepared from crop or weed residues reduced the severity of dry root rot in cluster beans caused by *M. phaseolina*. However, an increase in some diseases due to compost usage has also been observed, since compost is a product that varies considerably in chemical, physical, and biotic composition, and, consequently, also in its ability to suppress soil-borne diseases (Pugliese et al., 2015).

While compost application adds nutrients to the soil, most of them are in organic form and need to be mineralized before it becomes plant available. Thus, the gradual release of N from compost may reduce the likelihood of nitrate leaching. Lloyd et al. (2016) found that vermicompost and mushroom compost are good sources of early season nitrate for strawberry plants. If compost can supply sufficient nitrogen to the crop during the early season without triggering nitrate leaching, compost may be an effective alternative to synthetic slow-release fertilizers. Here we evaluated the effects of pre-plant compost application on N dynamics and suppression of *M. phaseolina* in strawberry cultivars with different levels of resistance to this soil-borne fungal pathogen.

8.1 Experimental design

A 2-year field trial was conducted to assess the effect of strawberry cultivar and preplant fertilizer management on nitrogen dynamics and disease incidence in a strawberry field infested with *M. phaseolina*. The experimental site was located at the California Polytechnic State University in San Luis Obispo, California (USA) on a silty clay loam soil. The experiment was laid as a randomized complete block design with four replications. Pre-plant fertilizers represented the main plots and included three treatments: a control with no pre-plant fertilizer applied (control), a treatment with AgRx brand 18-8-13 urea coated custom blended strawberry preplant controlled release fertilizer (CRF), and a treatment with preplant compost application (compost). CRF was applied at a rate equivalent to 112 kg N ha^{-1} representing typical grower practice. Dairy manure-based compost was incorporated through hand-tilling in the top 15 cm of soil and applied at a rate equivalent to 100 kg N ha^{-1}, equating to 10-12 t ha^{-1} accounting for year-to-year variation in compost moisture content. This rate also matched the total N input in the CRF treatment. This application rate equated to 1.2 t C ha^{-1} per dry weight basis (Fig. 2). It should be noted that only a fraction of this N is expected to be mineralized in the first year following compost application. All treatments received the same in-season fertilizer applications with conventional liquid fertilizer commonly used in CA strawberry production through the drip irrigation system. Each main plot was split into 4 subplots, each of which was planted with one of four strawberry cultivars: Albion, Monterey, San Andreas, and a proprietary variety. Cultivars were selected based on preliminary evidence suggesting differences in N requirements. Yield data was collected by hand-picking treatment plots twice per week during the harvesting season between April and July each year, representative of grower practice. Disease incidence was expressed as area under disease progress curve (AUDPC), a metric that

Figure 2 From left to right: Rhizon pore water sampler with porous ceramic tip installed to 30 cm depth (below root zone); preparation of strawberry fields includes shaping soil into raised beds with a tractor implement, a similar implement is used to apply synthetic pre-plant fertilizer; strawberry plants photographed in May 2020 showing symptoms of infection by *Macrophomina phaseolina* (brown plants).

considers the plant mortality rate (determined visually) as well as the timing of disease incidence within the growing season and is expressed as %-days (Jeger and Viljanen-Rollinson, 2001; Madden et al., 2017). Soil pore water samples were collected every two weeks (from April until July) from rhizon samplers (Rhizosphere Research Products (RRP), the Netherlands) installed at 15 cm and 30 cm depth to assess the concentration of nitrate in and below the root zone, respectively (Fig. 2). Soil pore water NO_3^--N concentration was converted to soil NO_3^--N concentration based on soil moisture content. To determine soil NO_3^--N exposure, the summed concentrations were plotted against time and trapezoidal- integration used to represent cumulative soil NO_3^--N availability across the entire analysis period. Linear-mixed effects analysis (ANOVA) was performed in R to assess the effects of factors on response variables. The Shapiro-Wilk test and Levene's test were performed to confirm that our data met the assumptions of normality and homogeneity of variance.

8.2 Results and discussion

There was a significant effect of pre-plant fertilizer treatment in *M. phaseolina* disease incidence in the 2019 and 2020 growing seasons ($P = 0.017$ and $P < .001$, respectively). Nevertheless, opposite to what we had initially hypothesized, compost application did not suppress disease incidence in this case study, but it increased disease incidence compared to the control treatment where no pre-plant fertilizer was applied in one of the growing seasons (Fig. 3). In 2019, disease incidence in the control treatment (169.8%-days) was significantly less than in the compost and CRF treatments (249.9%-days and 256.3%-days, respectively), and in 2020, disease incidence in the control and compost treatments (343.7%-days and 451.2%-days, respectively) were significantly less than in the CRF treatment (9863%-days). While various mechanisms by which

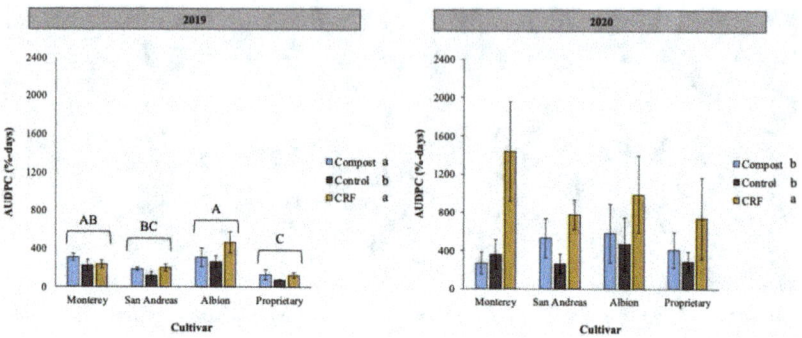

Figure 3 Disease incidence determined using the area under the disease progress curve (AUDPC) from observation data collected from May 10, 2019, to July 19, 2019 (year 1) and May 1, 2020, to July 24, 2020 (year 2). Error bars represent standard errors of the mean for each treatment ($n = 4$). Uppercase letters indicate significant differences in disease incidence between cultivar in 2019 measurements at $P < 0.05$. Lowercase letters indicate significant differences in disease incidence between pre-plant fertilizer treatments in 2019 and 2020 measurements at $P < 0.05$.

compost contributes to disease suppressiveness have been demonstrated, mixed effects of compost application on disease incidence have been observed (Mehta et al., 2014). Variations in effects of compost application on disease suppressiveness across studies have been attributed to variations in the cropping system, the compost application rate, method of application, frequency of compost amendments, compost feedstock and quality, and the compost microbiome (De Corato, 2020). For example, Tubeileh and Stephenson (2020) observed a reduction in the soil load of *Verticillium dahliae* in soils applied with 25 t ha^{-1} of composts derived from grape pomace, olive pomace-dairy manure mix, or mixed crop residue, but compost made of dairy and horse manure alone was much less effective at suppressing the pathogen. If compost application is aimed at suppressing disease, identifying a suitable compost source, rate, and application method is essential.

There was a significant effect of cultivar on total yield in both 2019 and 2020 ($P < 0.001$) (Fig. 4). There was no significant effect of pre-plant fertilizer on total yield in 2019, but there was a significant effect of pre-plant fertilizer on total yield in 2020 ($P = 0.041$) (Fig. 4). The compost and control treatment had a greater mean total yield in 2020 compared with the CRF treatment. Prior to planting, soil tests indicated a residual soil nitrate concentration of 75 mg kg^{-1} in 2019 and 21 mg kg^{-1} in 2020. Despite the stark differences in N applications between treatments, no effects of preplant fertilizer treatment on nitrate exposure in or below the root zone were observed during the 2019 or 2020 growing seasons (Fig. 5). Potentially, the release of nitrogen from both the compost and the CRF treatment was slow enough for plant uptake, nitrate immobilization, and denitrification to reduce the concentration of

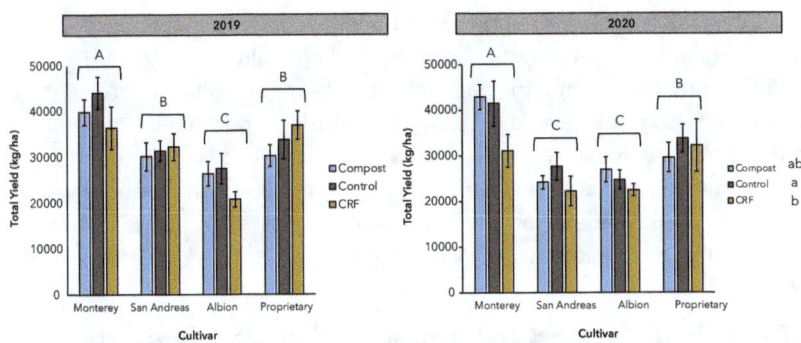

Figure 4 Total yield (kg ha⁻¹) from strawberry harvests collected from April 1, 2019, to July 19, 2019 (year 1) and April 4, 2020, to July 24, 2020 (year 2). Error bars represent standard errors of the mean for each treatment ($n = 4$). Uppercase letters indicate significant differences in total yield between cultivars in 2019 and 2020 measurements at $P < 0.05$. Lowercase letters indicate significant differences in total yield between pre-plant fertilizer treatments in 2020 measurements at $P < 0.05$.

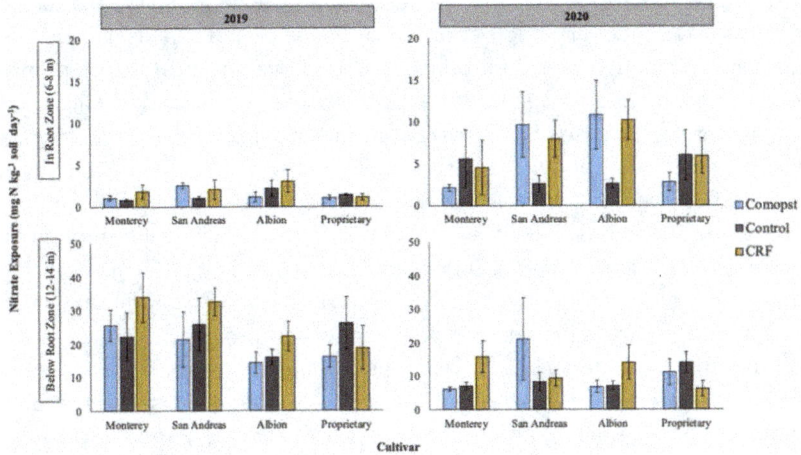

Figure 5 Soil nitrate exposure from January to July 2019 and November 2019 to April 2020 for the 2019 and 2020 growing seasons, respectively. Error bars represent standard errors of the mean for each treatment ($n = 4$).

NO_3^- in the soil. Tightening of N cycling by more heavily relying on microbial turnover of N between the soil organic matter pool and plant available forms has been identified as a key strategy to reduce N leaching from agricultural systems (Drinkwater and Snapp, 2007; Schimel and Bennett, 2004). Given that organic agriculture typically relies on compost amendments, comparison of organic versus conventional systems can be used as a proxy to assess the effect of compost on N loss. In a recent review, Seufert and Ramankutty (2017) established that there is great uncertainty in the impact of organic management on water quality but N leaching per unit of surface area tended to be lower in organic agriculture compared to conventional agriculture.

In summary, despite the lack of short-term benefits observed in this experiment, compost use did not reduce yields and should therefore be considered as a sustainable alternative to control released fertilizers, given the long-term benefits for soil health shown in the literature. More research is needed to better understand the impact of compost source, rate, timing, and placement on crop nutrition and N leaching.

9 The use of compost to improve soil health, sequester carbon and reduce greenhouse gas emissions in perennial crops: a case study in a Mediterranean vineyard

Soil health in vineyards around the world is threatened by intensive agricultural practices such as tillage and herbicide use, particularly in semi-arid Mediterranean regions (Salomé et al., 2016). Sustainable soil management practices have gradually been adopted by the wine industry in the past few decades to minimize soil disturbance and improve soil health. The primary focus on healthy soil management practices is usually on increasing soil organic matter content, because of its fundamental role in supporting many biological, physical, and chemical dynamic aspects of soil health (Liu et al., 2016; Rawls et al., 2003). Practices such as the use of cover crops, reduced till, and the application of compost have the potential to improve soil health through the increase in soil organic matter. Furthermore, by increasing soil organic matter, these practices lead to sequestration of C in soils, having a large potential for mitigation of climate change. Compost application provides a direct input of C and nutrients to the soil (Lal, 2004) which increases microbial biomass and promotes aggregate formation, facilitating C stabilization and sequestration as biomass or within soil aggregates (Six et al., 2004). Macroaggregates provide a preferential site for microaggregate formation and the associated C stabilization (Six et al., 2000), and long-term compost input was found to increase the fraction of large macroaggregate (Liu et al., 2021). Furthermore, compost can indirectly increase soil C storage by supplying nutrients and increasing above and belowground vine biomass and root exudation, a mechanism which has been recently found to contribute to most of the sequestered or stable C in soils. As woody perennials, grapevines have great potential for C sequestration and climate change mitigation under the right management practices (Wolff et al., 2018). On the other hand, the mineralization of labile organic C and N from compost application can lead to the subsequent release of carbon dioxide (CO_2) and nitrous oxide (N_2O) through increased microbial respiration, nitrification, and denitrification (Bustamante et al., 2011; Lazcano et al., 2013;

Verhoeven et al., 2019). This release of CO_2 and N_2O can offset the climate change mitigation benefits of compost, particularly in a situation where soil C sequestration potential has reached its plateau (Guenet et al., 2020; Lugato et al., 2018). Compost application rate is one of the main factors affecting the potential benefits for soil health (Martínez-Blanco et al., 2013b). While high application rates can increase nutrient and C inputs to the soil therefore having stronger benefits for soil health, they could also trigger large emissions of greenhouse gases, offsetting the benefits of compost for climate change mitigation. Here, we evaluated these potential tradeoffs by measuring the effects of increasing doses of compost application on soil C, CO_2, and N_2O emissions in a Mediterranean vineyard.

9.1 Experimental design and analyses

A 3-year study investigating the effects of compost application rate on soil C sequestration and greenhouse gas (GHG) emissions was conducted in a commercial wine grape vineyard (*Vitis vinifera*, var. Cabernet Sauvignon) located at J. Lohr Vineyards and Wines in the Paso Robles American Viticultural Area (AVA) near Paso Robles, California (USA) between November 2018 and November 2020. The experimental setup was a randomized completed block design with four treatment levels (compost application rates) replicated four times: 0 (control), 4.5, 9.0, and 13.5 Mg ha^{-1} year^{-1} (fresh weight, 0, 0.45, 0.91, and 1.36 Mg C ha^{-1} equivalent). Individual plots included four rows of 30 vines and three interrow areas (tractor rows), while the two middle vine rows and the middle tractor rows were the sampling areas, within which two functional locations (vine and tractor row) were designated for soil and GHG sampling. Compost (certified organic through California Department of Food and Agriculture's Organic Input Materials Program) made of processed livestock manure and green waste was broadcasted annually over the entire orchard floor after harvest when the rain (wet) season started in the region. During the experimental period, soil samples to 60 cm depth were collected in spring each year from the two functional locations. The effects of compost application rate on soil C sequestration were measured by soil health indicators including total soil C (%), permanganate oxidizable carbon (POXC), and aggregate size distribution. Total soil C was determined by combustion of approximately 1 g of air-dried, ground soil in a Vario Max CNS elemental analyzer (Elementar, Langenselbold, Hesse, Germany) at 900°C. POXC was colorimetrically determined by reacting air-dried soil samples with 2 M potassium permanganate solution using the revised protocol of Weil et al. (2003). Aggregate size distribution was done by wet-sieving based on the methodology described in Six et al. (2000), and four aggregate-size fractions were separated from this procedure: silt and clay (S+C; <53 µm), microaggregate (m; 53-250 µm), small macroaggregate (SM;

250-2000 µm), and large macroaggregate (LM; >2000 µm). GHG samples were also collected from both functional locations to measure cumulative CO_2 and N_2O emissions using the static chamber technique (Hutchinson and Mosier, 1981; Parkin and Venterea, 2010) and USDA GraceNet guidelines for static chamber design and GHG sampling procedures (Fig. 6). The seasons were arbitrarily defined: dry season started on the first day of cover crop mowing in early to mid-April, and the wet season began after harvest when the associated N_2O fluxes had subsided to background level. Linear mixed effects analysis (ANOVA) was performed in R to assess the effects of factors on response variables. The Shapiro-Wilk test and Levene's test were performed to confirm that our data met the assumptions of normality and homogeneity of variance.

9.2 Results and discussion

After 2 years of compost application, total soil C content did not increase significantly when compost input increased (Table 2). This is likely due to the slow C turnover rates in soil. However, POXC, which represents the active C pool that is sensitive to soil management and tends to be stabilized in soil (Hurisso et al., 2016), was significantly increased by the rate of compost application (Fig. 7). Higher compost application rates resulted in higher concentrations of POXC to a depth of 60 cm ($P < 0.001$). This suggests that annually applying compost up to the rate of 13.5 Mg ha^{-1} in a vineyard can increase the tendency of C stabilization and potential sequestration in soil. In addition, compost application rate also had a significant interactive effect on the distribution of the large macroaggregate (LM) fraction of soil (Fig. 8). Significantly more LM was found in the tractor row topsoil than the subsoil and both depth increments under the vine ($P < 0.001$). The LM distribution within the tractor row topsoil was marginally and significantly greater at compost application rates of 9.0 and 13.5 Mg ha^{-1} than the control treatment ($P = 0.0755$ at 90% confidence interval), suggesting an early sign towards increased macroaggregate

Figure 6 From left to right: broadcasted of compost at the vineyard and placement of the static chambers used to sample GHG at the two functional locations within each experimental plot.

Table 2 C content (Mg C ha^{-1}) and the respective standard errors of the means ($n = 4$) by compost application rate, functional location, and depth in April 2020, after two annual compost applications.

Rate (t ha^{-1})	0-15 cm		15-30 cm		30-60 cm	
	Tractor row	Vine row	Tractor row	Vine row	Tractor row	Vine row
0	15.3 ± 0.48	10.1 ± 0.61	6.3 ± 0.23	6.0 ± 0.23	10.5 ± 0.54	10.3 ± 0.43
4.5	16.9 ± 1.46	7.2 ± 0.41	6.9 ± 0.50	6.0 ± 0.24	9.7 ± 1.14	8.3 ± 0.99
9.0	17.5 ± 1.05	8.4 ± 0.34	6.7 ± 0.58	6.4 ± 0.53	9.4 ± 0.70	9.2 ± 0.37
13.5	18.9 ± 1.61	9.4 ± 1.35	6.2 ± 0.13	6.7 ± 0.19	9.2 ± 0.18	9.8 ± 0.57
Location by depth interaction						
Tractor row	17.2 ± 1.29a		6.5 ± 0.39d		9.7 ± 0.69c	
Vine row	9.3 ± 0.76b		6.3 ± 0.33d		9.4 ± 0.68cd	

Lower case letters indicate significant differences between locations and depths at $P < 0.05$.

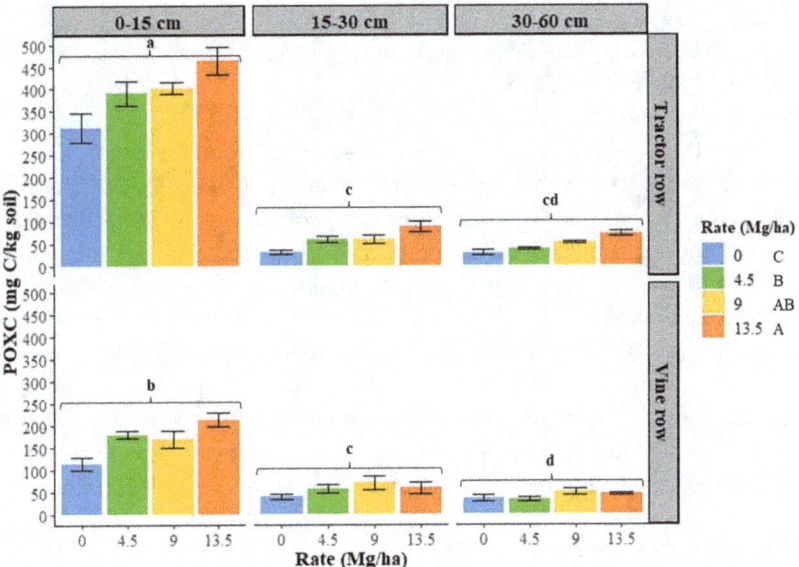

Figure 7 POXC (mg C kg^{-1} soil) by depth and functional location measured in April 2020, after two annual compost applications. Error bars represent standard errors of the means ($n = 4$). Lowercase letters indicate significant differences between depths and locations at $P < 0.05$. Upper case letters indicate significant differences between compost application rates at $P < 0.05$.

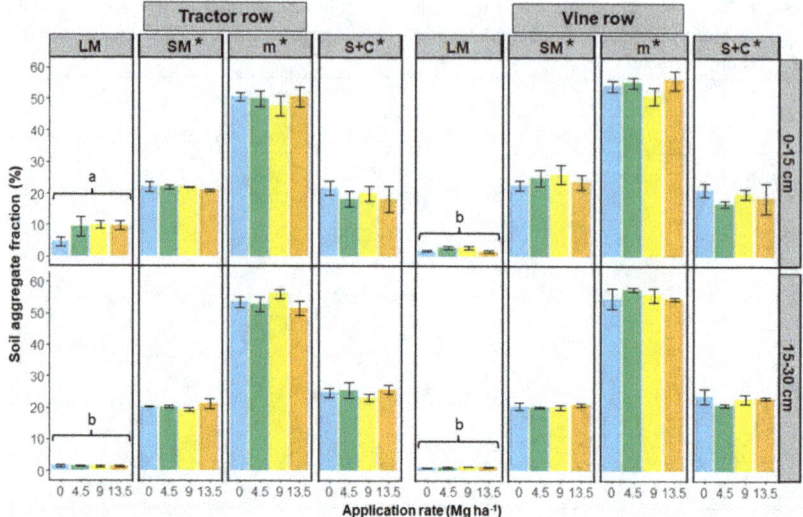

Figure 8 Fractions (%) of soil aggregates by aggregate size and functional location at the depths of 0–15 and 15–30 cm in April 2020. LM, SM, m, and S+C represent the four aggregate size categories: large macroaggregate, small macroaggregate, microaggregate, and silt and clay, respectively. Error bars are standard errors of the means (n = 4). Lowercase letters indicate significant differences between locations and depths, while asterisks (*) indicate significant differences between depths at $P < 0.05$ within each aggregate size category.

formation through biological activity boosted by the addition of compost, and further C stabilization within these macroaggregates (Liu et al., 2021; Six et al., 2004). When it comes to GHG emissions, both cumulative CO_2 (Fig. 9) and N_2O (Fig. 10) emissions at both functional locations measured in two wet (winter-spring) and two dry (summer-fall) seasons were not significantly affected by compost application. This lack of significant effect of compost addition and the application rates was a welcomed finding that suggests the environmental tradeoff in the form of CO_2 and N_2O emissions from the use of compost up to 6 tons/acre may be minimal in vineyards. Similar to GHG emissions, grape yield was also not significantly affected by compost application in this study (Table 3).

In conclusion, although no significant changes in total soil C content were observed, applying compost annually up to 6 tons/acre was shown to have positive effects on early signs of C sequestration, as indicated by increased POXC and large macroaggregate formation. Long-term commitment to this practice will be required to show the desired increase in total soil C content. Since compost application did not significantly increase cumulative GHG emissions in this study, nor have detrimental effects on grape yield, it should be recommended to wine growers for the purpose of conserving soil organic matter and maintaining soil health.

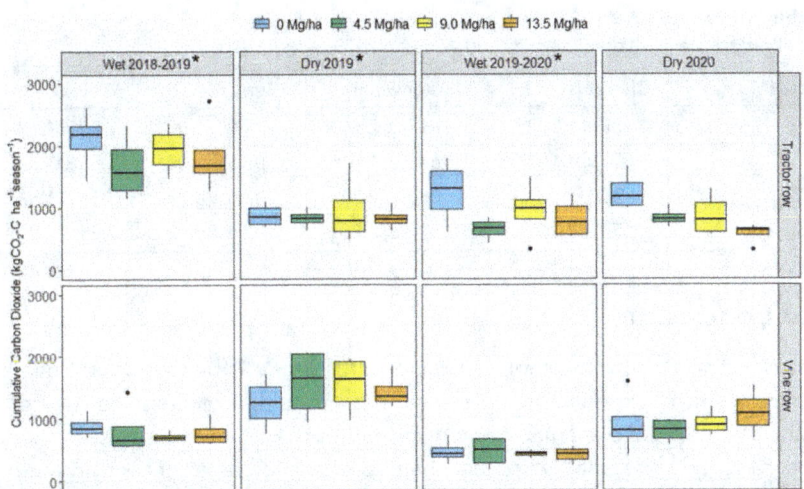

Figure 9 Cumulative CO_2 emissions measured in soil by functional location under compost application treatments at rates of 0 Mg ha^{-1}, 4.5 Mg ha^{-1}, 9.0 Mg ha^{-1}, and 13.5 Mg ha^{-1} during the two wet and two dry seasons from 2018 to 2020. Error bars represent standard errors of the mean ($n = 4$). Asterisks (*) indicate significant differences between locations within the same season and application rate at $P < 0.05$.

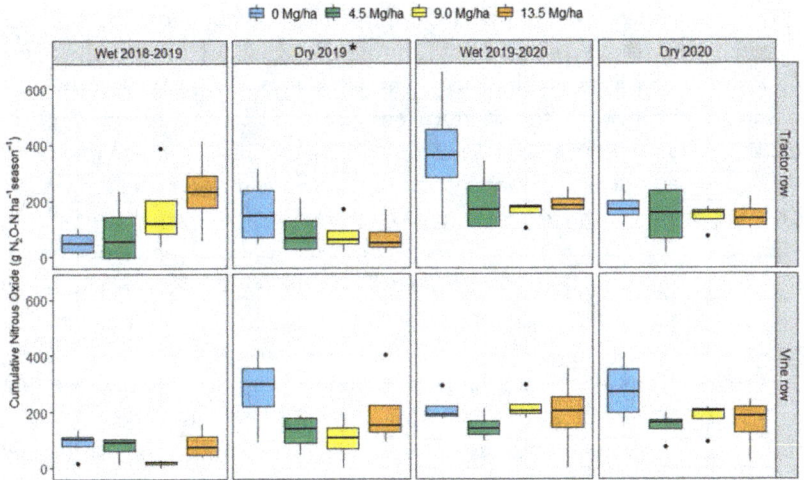

Figure 10 Cumulative N_2O emissions measured in soil by functional location under compost application treatments at rates of 0 Mg ha^{-1}, 4.5 Mg ha^{-1}, 9.0 Mg ha^{-1}, and 13.5 Mg ha^{-1} during the two wet and two dry seasons from 2018 to 2020. Error bars represent standard errors of the mean ($n = 4$). Asterisk (*) indicate significant differences between locations at $P < 0.05$ within the dry season.

Table 3 Grape yield components and standard errors of the means (*n* = 4) by compost application rate in J. Lohr Vineyards and Wines in 2020

Rate (Mg ha⁻¹)	Clusters vine⁻¹	Cluster weight (g)	Berry mass (g)	Yield (t ha⁻¹)
0	52 ± 1.5	69.8 ± 2.60	0.69 ± 0.03	7.27 ± 1.63
4.5	50 ± 4.1	64.2 ± 1.41	0.75 ± 0.02	6.56 ± 1.85
9.0	58 ± 5.6	68.7 ± 2.34	0.70 ± 0.03	7.96 ± 1.58
13.5	54 ± 3.5	67.3 ± 3.25	0.68 ± 0.02	7.23 ± 2.01

10 Conclusion

Compost application has short-, medium-, and long-term benefits for soil health. There is strong experimental evidence that shows that compost constitutes a valuable source of plant macro and micronutrients. Furthermore, these nutrients are released gradually which can help prevent losses to the environment. The supply of organic matter boosts microbial biomass and activity, shifting the composition of the microbial community. Increases in microbial biomass and organic matter inputs lead to sequestration of C which can help mitigate climate change. Nevertheless, the observed effects were extremely variable. Furthermore, evidence showing the effects of compost on crop yield, quality, and health is weaker, ranging from positive to negative effects.

Evidence shows that compost can impede plant growth due to the supply of excessive amounts of salts or phytotoxic compounds. There are also environmental risks associated with the use of compost such as the excess application of easily available nitrogen and P which can lead to eutrophication of water courses when transported through leaching and runoff, or to the emission of greenhouse gasses such as nitrous oxide. Certain composts can carry excess amounts of contaminants such as heavy metals or impurities such as microplastics which are known to have clear negative consequences for the structure and function of the soil ecosystem. Finally, the potential presence of enteric pathogens such as *Salmonella* or *E. coli* poses a serious risk to human health, particularly when compost is applied to crops that will be consumed fresh.

Thus, even though there are clear benefits to the use of compost, there are also certain risks which can be avoided by understanding the factors that drive this large variability in agronomic, environmental, and health outcomes. Most risks can be avoided through the adequate management of the composting process. Particularly, the thermophilic stage is critical for the elimination of phytotoxic substances and enteric pathogens and therefore for compost stabilization. Selecting high-quality feedstocks, free of impurities and contaminants, contributes to increasing compost quality and therefore can maximize the benefits of compost to soil health while decreasing environmental risks.

Variability in compost agronomic effects can be attributed to different crop requirements, compost management, compost quality, soil properties, and environmental conditions. In terms of management, the number and rate of applications strongly influence the outcomes for soil health. Repeated and large application rates lead to stronger beneficial effects in the soil health indicators evaluated but can also lead to the buildup of easily available N and P and therefore the associated risk for environmental pollution. Compost C:N ratio or particle size also drives the magnitude of the effects by regulating the rate of nutrient release. Soil properties such as texture have a strong effect on organic matter mineralization rates and C sequestration after compost application. In terms of environmental properties, seasonality, average temperatures, and annual precipitation rates were revealed to be important in driving the magnitude of the effects of compost on soil health.

Altogether, this evidences the need for site-specific knowledge and management in order to maximize compost benefits and reduce risks. Selecting high-quality feedstocks, managing the composting process to allow for the stabilization, sanitization, and curing of the organic material, understanding the properties of the finished compost, adjusting compost application rates and times to match crop needs are all required steps to maximize the benefits of compost for soil health. Furthermore, the potential benefits or risks of compost need to be considered within the context of other management strategies that may affect its performance, such as irrigation, the use of cover crops, or soil tillage, among others. Finally, evidence showing long-term effects of compost is particularly scarce; there is a strong need for standardized and reproducible long-term research that disentangles the interactions between compost, soil biota, and plants, across different soils and environmental conditions.

11 Where to look for further information

The journal 'Compost Science and Utilization' is a quarterly peer-reviewed journal that reports on advancements and breakthroughs in the science and engineering of compost production, compost product quality, and the utilization of composted materials.

The Test Method for the Examination of Composting and Compost (TMECC) of the US Composting Council, provides detailed protocols for the composting industry to verify the physical, chemical, and biological condition of composting feedstocks, material in process and compost products at the point of sale.

The European Compost Network, provides information on European policies on the use of compost and protection of soil health. https://www .compostnetwork.info.

Current Approaches and Future Trends in Compost Quality Criteria for Agronomic, Environmental and Human Health Benefits by Maria Pilar Bernal, Sven G. Sommer, Dave Chadwick, Chen Qing, Li Guoxue and Frederick C. Michel Jr. 2017, Advances in Agronomy.

12 References

Abiven, S., Menasseri, S. and Chenu, C. 2009. The effects of organic inputs over time on soil aggregate stability - a literature analysis. *Soil Biology and Biochemistry* 41(1), 1-12. https://doi.org/10.1016/j.soilbio.2008.09.015.

Aggelides, S. M. and Londra, P. A. 2000. Effects of compost produced from town wastes and sewage sludge on the physical properties of a loamy and a clay soil. *Bioresource Technology* 71(3), 253-259. https://doi.org/10.1016/S0960-8524(99)00074-7.

Al-Suhaibani, N., Selim, M., Alderfasi, A. and El-Hendawy, S. 2020. Comparative performance of integrated nutrient management between composted agricultural wastes, chemical fertilizers, and biofertilizers in improving soil quantitative and qualitative properties and crop yields under arid conditions. *Agronomy* 10(10), 1503. https://doi.org/10.3390/agronomy10101503.

Antille, D. L., Sakrabani, R. and Godwin, R. J. 2014. Phosphorus release characteristics from biosolids-derived organomineral fertilizers. *Communications in Soil Science and Plant Analysis* 45(19), 2565-2576. https://doi.org/10.1080/00103624.2014.912300.

Arthur, E., Cornelis, W. and Razzaghi, F. 2012. Compost amendment to sandy soil affects soil properties and greenhouse tomato productivity. *Compost Science and Utilization* 20(4), 215-221. https://doi.org/10.1080/1065657X.2012.10737051.

Arthur, E., Cornelis, W. M., Vermang, J. and De Rocker, E. 2011. Amending a loamy sand with three compost types: impact on soil quality: compost effect on the soil quality of loamy sand. *Soil Use and Management* 27(1), 116-123. https://doi.org/10.1111/j .1475-2743.2010.00319.x.

Avilés, M. and Borrero, C. 2017. Identifying characteristics of Verticillium Wilt suppressiveness in olive mill composts. *Plant Disease* 101(9), 1568-1577. https://doi .org/10.1094/PDIS-08-16-1172-RE.

Ba, S., Qu, Q., Zhang, K. and Groot, J. C. J. 2020. Meta-analysis of greenhouse gas and ammonia emissions from dairy manure composting. *Biosystems Engineering* 193, 126-137. https://doi.org/10.1016/j.biosystemseng.2020.02.015.

Bai, Z., Ma, W., Ma, L., Velthof, G. L., Wei, Z., Havlík, P., Oenema, O., Lee, M. R. F. and Zhang, F. 2018. China's livestock transition: driving forces, impacts, and consequences. *Science Advances* 4(7), eaar8534. https://doi.org/10.1126/sciadv.aar8534.

Bardgett, R. D. and van der Putten, W. H. 2014. Belowground biodiversity and ecosystem functioning. *Nature* 515(7528), 505-511. https://doi.org/10.1038/nature13855.

Bastida, F., Kandeler, E., Moreno, J. L., Ros, M., García, C. and Hernández, T. 2008. Application of fresh and composted organic wastes modifies structure, size and activity of soil microbial community under semiarid climate. *Applied Soil Ecology* 40(2), 318-329. https://doi.org/10.1016/j.apsoil.2008.05.007.

Bastida, F., Moreno, J. L., García, C. and Hernández, T. 2007. Addition of urban waste to semiarid degraded soil: long-term effect. *Pedosphere* 17(5), 557–567. https://doi .org/10.1016/S1002-0160(07)60066-6.

Bedada, W., Karltun, E., Lemenih, M. and Tolera, M. 2014. Long-term addition of compost and NP fertilizer increases crop yield and improves soil quality in experiments on smallholder farms. *Agriculture, Ecosystems and Environment* 195, 193–201. https:// doi.org/10.1016/j.agee.2014.06.017.

Bernal, M. P., Alburquerque, J. A. and Moral, R. 2009. Composting of animal manures and chemical criteria for compost maturity assessment. A review. *Bioresource Technology* 100(22), 5444–5453. https://doi.org/10.1016/j.biortech.2008.11.027.

Bernal, M. P., Sommer, S. G., Chadwick, D., Qing, C., Guoxue, L. and Michel, F. C. 2017. Current approaches and future trends in compost quality criteria for agronomic, environmental, and human health benefits. In: *Advances in Agronomy*. Elsevier, pp. 143–233. https://doi.org/10.1016/bs.agron.2017.03.002.

Bläsing, M. and Amelung, W. 2018. Plastics in soil: analytical methods and possible sources. *Science of the Total Environment* 612, 422–435. https://doi.org/10.1016/j .scitotenv.2017.08.086.

Boldrin, A., Andersen, J. K., Møller, J., Christensen, T. H. and Favoino, E. 2009. Composting and compost utilization: accounting of greenhouse gases and global warming contributions. *Waste Management and Research* 27(8), 800–812. https://doi.org/10 .1177/0734242X09345275.

Bonanomi, G., Antignani, V., Capodilupo, M. and Scala, F. 2010. Identifying the characteristics of organic soil amendments that suppress soilborne plant diseases. *Soil Biology and Biochemistry* 42(2), 136–144. https://doi.org/10.1016/j.soilbio .2009.10.012.

Bonanomi, G., Antignani, V., Pane, C. and Scala, F. 2007. Suppression of soilborne fungal diseases with organic amendments. *Journal of Plant Pathology* 89, 311–324.

Bottoms, T. G., Bolda, M. P., Gaskell, M. L. and Hartz, T. K. 2013a. Determination of strawberry nutrient optimum ranges through diagnosis and recommendation integrated system analysis. *HortTechnology* 23(3), 312–318. https://doi.org/10 .21273/HORTTECH.23.3.312.

Bottoms, T. G., Hartz, T. K., Cahn, M. D. and Farrara, B. F. 2013b. Crop and soil nitrogen dynamics in annual strawberry production in California. *HortScience* 48(8), 1034–1039. https://doi.org/10.21273/HORTSCI.48.8.1034.

Brady, N. C. and Weil, R. R. 2016. *The Nature and Properties of Soils* (15th ed.). Pearson, Columbus.

Brandt, K., Leifert, C., Sanderson, R. and Seal, C. J. 2011. Agroecosystem management and nutritional quality of plant foods: the case of organic fruits and vegetables. *Critical Reviews in Plant Sciences* 30(1–2), 177–197. https://doi.org/10.1080/07352689 .2011.554417.

Bronick, C. J. and Lal, R. 2005. Soil structure and management: a review. *Geoderma* 124(1–2), 3–22. https://doi.org/10.1016/j.geoderma.2004.03.005.

Bünemann, E. K., Bongiorno, G., Bai, Z., Creamer, R. E., De Deyn, G., de Goede, R., Fleskens, L., Geissen, V., Kuyper, T. W., Mäder, P., Pulleman, M., Sukkel, W., van Groenigen, J. W. and Brussaard, L. 2018. Soil quality – a critical review. *Soil Biology and Biochemistry* 120, 105–125. https://doi.org/10.1016/j.soilbio.2018.01.030.

Bustamante, M. A., Alburquerque, J. A., Restrepo, A. P., de la Fuente, C., Paredes, C., Moral, R. and Bernal, M. P. 2012. Co-composting of the solid fraction of anaerobic digestates, to obtain added-value materials for use in agriculture. *Biomass and Bioenergy* 43, 26–35. https://doi.org/10.1016/j.biombioe.2012.04.010.

Bustamante, M. A., Paredes, C., Marhuenda-Egea, F. C., Pérez-Espinosa, A., Bernal, M. P. and Moral, R. 2008. Co-composting of distillery wastes with animal manures: carbon and nitrogen transformations in the evaluation of compost stability. *Chemosphere* 72(4), 551–557. https://doi.org/10.1016/j.chemosphere.2008.03 .030.

Bustamante, M. A., Said-Pullicino, D., Agulló, E., Andreu, J., Paredes, C. and Moral, R. 2011. Application of winery and distillery waste composts to a Jumilla (SE Spain) vineyard: effects on the characteristics of a calcareous sandy-loam soil. *Agriculture, Ecosystems and Environment* 140(1-2), 80–87. https://doi.org/10.1016/j.agee.2010 .11.014.

Butler, T. J. and Muir, J. P. 2006. Dairy manure compost improves soil and increases tall wheatgrass yield. *Agronomy Journal* 98(4), 1090–1096. https://doi.org/10.2134/ agronj2005.0348.

Butterbach-Bahl, K., Baggs, E. M., Dannenmann, M., Kiese, R. and Zechmeister-Boltenstern, S. 2013. Nitrous oxide emissions from soils: how well do we understand the processes and their controls? *Philosophical Transactions of the Royal Society of London. Series B, Biological Sciences* 368(1621), 20130122. https://doi.org/10.1098 /rstb.2013.0122.

Calbrix, R., Barray, S., Chabrerie, O., Fourrie, L. and Laval, K. 2007. Impact of organic amendments on the dynamics of soil microbial biomass and bacterial communities in cultivated land. *Applied Soil Ecology* 35(3), 511–522. https://doi.org/10.1016/j .apsoil.2006.10.007.

Calleja-Cervantes, M. E., Fernández-González, A. J., Irigoyen, I., Fernández-López, M., Aparicio-Tejo, P. M. and Menéndez, S. 2015. Thirteen years of continued application of composted organic wastes in a vineyard modify soil quality characteristics. *Soil Biology and Biochemistry* 90, 241–254. https://doi.org/10.1016/j.soilbio.2015.07 .002.

Carle, S. F., Esser, B. K. and Moran, J. E. 2006. High-resolution simulation of basin-scale nitrate transport considering aquifer system heterogeneity. *Geosphere* 2(4), 195. https://doi.org/10.1130/GES00032.1.

Carvalhais, N., Forkel, M., Khomik, M., Bellarby, J., Jung, M., Migliavacca, M., ⊡u, M., Saatchi, S., Santoro, M., Thurner, M., Weber, U., Ahrens, B., Beer, C., Cescatti, A., Randerson, J. T. and Reichstein, M. 2014. Global covariation of carbon turnover times with climate in terrestrial ecosystems. *Nature* 514, 213–217. https://doi.org/10 .1038/nature13731.

Castán, E., Satti, P., González-Polo, M., Iglesias, M. C. and Mazzarino, M. J. 2016. Managing the value of composts as organic amendments and fertilizers in sandy soils. *Agriculture, Ecosystems and Environment* 224, 29–38. https://doi.org/10.1016/j .agee.2016.03.016.

Celestina, C., Hunt, J. R., Sale, P. W. G. and Franks, A. E. 2019. Attribution of crop yield responses to application of organic amendments: a critical review. *Soil and Tillage Research* 186, 135–145. https://doi.org/10.1016/j.still.2018.10.002.

Celik, I., Gunal, H., Budak, M. and Akpinar, C. 2010. Effects of long-term organic and mineral fertilizers on bulk density and penetration resistance in semi-arid Mediterranean soil conditions. *Geoderma* 160(2), 236–243. https://doi.org/10.1016/j.geoderma.2010 .09.028.

Chadwick, D., Wei, J., Yan'an, T., Guanghui, Y., Qirong, S. and Qing, C. 2015. Improving manure nutrient management towards sustainable agricultural intensification in China. *Agriculture, Ecosystems and Environment* 209, 34–46. https://doi.org/10.1016/j.agee.2015.03.025.

Chalk, P. M., Magalhães, A. M. T. and Inácio, C. T. 2013. Towards an understanding of the dynamics of compost N in the soil-plant-atmosphere system using 15N tracer. *Plant and Soil* 362(1–2), 373–388. https://doi.org/10.1007/s11104-012-1358-5.

Chan, K. Y., Fahey, D. J., Newell, M. and Barchia, I. 2010. Using composted mulch in vineyards–effects on grape yield and quality. *International Journal of Fruit Science* 10(4), 441–453. https://doi.org/10.1080/15538362.2010.530135.

Charles, A., Rochette, P., Whalen, J. K., Angers, D. A., Chantigny, M. H. and Bertrand, N. 2017. Global nitrous oxide emission factors from agricultural soils after addition of organic amendments: a meta-analysis. *Agriculture, Ecosystems and Environment* 236, 88–98. https://doi.org/10.1016/j.agee.2016.11.021.

Chellemi, D. O. 2002. Nonchemical management of soilborne pests in fresh market vegetable production systems. *Phytopathology* 92(12), 1367–1372. https://doi.org/10.1094/PHYTO.2002.92.12.1367.

Chen, X. P., Zhang, Y. Q., Tong, Y. P., Xue, Y. F., Liu, D. Y., Zhang, W., Deng, Y., Meng, Q. F., Yue, S. C., Yan, P., Cui, Z. L., Shi, X. J., Guo, S. W., Sun, Y. X., Ye, Y. L., Wang, Z. H., Jia, L. L., Ma, W. Q., He, M. R., Zhang, X. Y., Kou, C. L., Li, Y. T., Tan, D. S., Cakmak, I., Zhang, F. S. and Zou, C. Q. 2017. Harvesting more grain zinc of wheat for human health. *Scientific Reports* 7(1), 7016. https://doi.org/10.1038/s41598-017-07484-2.

Chen, Y., Camps-Arbestain, M., Shen, Q., Singh, B. and Cayuela, M. L. 2018. The long-term role of organic amendments in building soil nutrient fertility: a meta-analysis and review. *Nutrient Cycling in Agroecosystems* 111(2–3), 103–125. https://doi.org/10.1007/s10705-017-9903-5.

Cherif, H., Ayari, F., Ouzari, H., Marzorati, M., Brusetti, L., Jedidi, N., Hassen, A. and Daffonchio, D. 2009. Effects of municipal solid waste compost, farmyard manure and chemical fertilizers on wheat growth, soil composition and soil bacterial characteristics under Tunisian arid climate. *European Journal of Soil Biology* 45(2), 138–145. https://doi.org/10.1016/j.ejsobi.2008.11.003.

Choi, W.-J., Jin, S.-A., Lee, S.-M., Ro, H.-M. and Yoo, S.-H. 2001. Corn uptake and microbial immobilization of 15N-labeled urea-N in soil as affected by composted pig manure. *Plant and Soil* 235(1), 1–9. https://doi.org/10.1023/A:1011896912888.

Colvero, D. A., Ramalho, J., Gomes, A. P. D., Matos, M. A. A. and Tarelho, L. A. D. C. 2020. Economic analysis of a shared municipal solid waste management facility in a metropolitan region. *Waste Management* 102, 823–837. https://doi.org/10.1016/j.wasman.2019.11.033.

Coria-Cayupán, Y. S., Sánchez de Pinto, M. I. and Nazareno, M. A. 2009. Variations in bioactive substance contents and crop yields of lettuce (Lactuca sativa L.) cultivated in soils with different fertilization treatments. *Journal of Agricultural and Food Chemistry* 57(21), 10122–10129. https://doi.org/10.1021/jf903019d.

Courtney, R. G. and Mullen, G. J. 2008. Soil quality and barley growth as influenced by the land application of two compost types. *Bioresource Technology* 99(8), 2913–2918. https://doi.org/10.1016/j.biortech.2007.06.034.

Crecchio, C., Curci, M., Mininni, R., Ricciuti, P. and Ruggiero, P. 2001. Short-term effects of municipal solid waste compost amendments on soil carbon and nitrogen content, some enzyme activities and genetic diversity. *Biology and Fertility of Soils* 34(5), 311–318. https://doi.org/10.1007/s003740100413.

Daynes, C. N., Field, D. J., Saleeba, J. A., Cole, M. A. and McGee, P. A. 2013. Development and stabilisation of soil structure via interactions between organic matter, arbuscular mycorrhizal fungi and plant roots. *Soil Biology and Biochemistry* 57, 683–694. https://doi.org/10.1016/j.soilbio.2012.09.020.

de Araújo, A. S. F., de Melo, W. J. and Singh, R. P. 2010. Municipal solid waste compost amendment in agricultural soil: changes in soil microbial biomass. *Reviews in Environmental Science and Bio/Technology* 9(1), 41–49. https://doi.org/10.1007/s11157-009-9179-6.

De Corato, U. 2020. Agricultural waste recycling in horticultural intensive farming systems by on-farm composting and compost-based tea application improves soil quality and plant health: a review under the perspective of a circular economy. *Science of the Total Environment* 738, 139840. https://doi.org/10.1016/j.scitotenv.2020.139840.

de Vries, F. T., Thebault, E., Liiri, M., Birkhofer, K., Tsiafouli, M. A., Bjornlund, L., Bracht Jorgensen, H., Brady, M. V., Christensen, S., de Ruiter, P. C., d'Hertefeldt, T., Frouz, J., Hedlund, K., Hemerik, L., Hol, W. H. G., Hotes, S., Mortimer, S. R., Setala, H., Sgardelis, S. P., Uteseny, K., van der Putten, W. H., Wolters, V. and Bardgett, R. D. 2013. Soil food web properties explain ecosystem services across European land use systems. *Proceedings of the National Academy of Sciences of the United States of America* 110(35), 14296–14301. https://doi.org/10.1073/pnas.1305198110.

Debertoldi, M., Vallini, G. and Pera, A. 1983. The biology of composting: a review. *Waste Management and Research* 1(2), 157–176. https://doi.org/10.1016/0734-242X(83)90055-1.

Degens, B. P. and Harris, J. A. 1997. Development of a physiological approach to measuring the catabolic diversity of soil microbial communities. *Soil Biology and Biochemistry* 29(9–10), 1309–1320. https://doi.org/10.1016/S0038-0717(97)00076-X.

Diacono, M. and Montemurro, F. 2010. Long-term effects of organic amendments on soil fertility. A review. *Agronomy for Sustainable Development* 30(2), 401–422. https://doi.org/10.1051/agro/2009040.

Diana, G., Beni, C. and Marconi, S. 2008. Organic and mineral fertilization: effects on physical characteristics and boron dynamic in an agricultural soil. *Communications in Soil Science and Plant Analysis* 39(9–10), 1332–1351. https://doi.org/10.1080/00103620802004037.

Domínguez, M., Paradelo Núñez, R., Piñeiro, J. and Barral, M. T. 2019. Physicochemical and biochemical properties of an acid soil under potato culture amended with municipal solid waste compost. *International Journal of Recycling of Organic Waste in Agriculture* 8(2), 171–178. https://doi.org/10.1007/s40093-019-0246-x.

Drinkwater, L. E. and Snapp, S. S. 2007. Nutrients in agroecosystems: rethinking the management paradigm. In: *Advances in Agronomy*. Elsevier, pp. 163–186. https://doi.org/10.1016/S0065-2113(04)92003-2.

Duong, T. T. T., Verma, S. L., Penfold, C. and Marschner, P. 2013. Nutrient release from composts into the surrounding soil. *Geoderma* 195–196, 42–47. https://doi.org/10.1016/j.geoderma.2012.11.010.

Eden, M., Gerke, H. H. and Houot, S. 2017. Organic waste recycling in agriculture and related effects on soil water retention and plant available water: a review. *Agronomy* for *Sustainable Development* 37(2), 11. https://doi.org/10.1007/s13593-017-0419 -9.

Eghball, B. 2003. Leaching of phosphorus fractions following manure or compost application. *Communications in Soil Science and Plant Analysis* 34(19–20), 2803–2815. https://doi.org/10.1081/CSS-120025207.

Eghball, B. and Power, J. F. 1999. Phosphorus- and nitrogen-based manure and compost applications corn production and soil phosphorus. *Soil Science Society of America Journal* 63(4), 895–901. https://doi.org/10.2136/sssaj1999.634895x.

Epelde, L., Jauregi, L., Urra, J., Ibarretxe, L., Romo, J., Goikoetxea, I. and Garbisu, C. 2018. Characterization of composted organic amendments for agricultural use. *Frontiers in Sustainable Food Systems* 2, 44. https://doi.org/10.3389/fsufs.2018 .00044.

Erhart, E., Hartl, W. and Putz, B. 2005. Biowaste compost affects yield, nitrogen supply during the vegetation period and crop quality of agricultural crops. *European Journal of Agronomy* 23(3), 305–314. https://doi.org/10.1016/j.eja.2005.01.002.

Erktan, A., Or, D. and Scheu, S. 2020. The physical structure of soil: determinant and consequence of trophic interactions. *Soil Biology and Biochemistry* 148, 107876. https://doi.org/10.1016/j.soilbio.2020.107876.

Fagnano, M., Adamo, P., Zampella, M. and Fiorentino, N. 2011. Environmental and agronomic impact of fertilization with composted organic fraction from municipal solid waste: a case study in the region of Naples, Italy. *Agriculture, Ecosystems and Environment* 141(1–2), 100–107. https://doi.org/10.1016/j.agee.2011.02.019.

Farrell, M., Griffith, G. W., Hobbs, P. J., Perkins, W. T. and Jones, D. L. 2010. Microbial diversity and activity are increased by compost amendment of metal-contaminated soil. *FEMS Microbiology Ecology* 71(1), 94–105. https://doi.org/10.1111/j.1574 -6941.2009.00793.x.

Fierer, N., Wood, S. A. and Bueno de Mesquita, C. P. 2021. How microbes can, and cannot, be used to assess soil health. *Soil Biology and Biochemistry* 153, 108111. https://doi .org/10.1016/j.soilbio.2020.108111.

Franco-Otero, V. G., Soler-Rovira, P., Hernández, D., López-de-Sá, E. G. and Plaza, C. 2012. Short-term effects of organic municipal wastes on wheat yield, microbial biomass, microbial activity, and chemical properties of soil. *Biology and Fertility of Soils* 48(2), 205–216. https://doi.org/10.1007/s00374-011-0620-y.

Frey, S. D., Lee, J., Melillo, J. M. and Six, J. 2013. The temperature response of soil microbial efficiency and its feedback to climate. *Nature Climate Change* 3(4), 395–398. https:// doi.org/10.1038/nclimate1796.

Frostegård, Å., Bååth, E. and Tunlio, A. 1993. Shifts in the structure of soil microbial communities in limed forests as revealed by phospholipid fatty acid analysis. *Soil Biology and Biochemistry* 25(6), 723–730. https://doi.org/10.1016/0038 -0717(93)90113-P.

García-Gil, J. C., Ceppi, S. B., Velasco, M. I., Polo, A. and Senesi, N. 2004. Long-term effects of amendment with municipal solid waste compost on the elemental and acidic functional group composition and pH-buffer capacity of soil humic acids. *Geoderma* 121(1–2), 135–142. https://doi.org/10.1016/j.geoderma.2003.11.004.

García-Gil, J. C., Plaza, C., Soler-Rovira, P. and Polo, A. 2000. Long-term effects of municipal solid waste compost application on soil enzyme activities and microbial biomass. *Soil Biology and Biochemistry* 32(13), 1907–1913. https://doi.org/10.1016 /S0038-0717(00)00165-6.

Garland, J. L. and Mills, A. L. 1991. Classification and characterization of heterotrophic microbial communities on the basis of patterns of community-level sole-carbon-source utilization. *Applied and Environmental Microbiology* 57(8), 2351–2359. https://doi.org/10.1128/aem.57.8.2351-2359.1991.

Giannakis, G. V., Kourgialas, N. N., Paranychianakis, N. V., Nikolaidis, N. P. and Kalogerakis, N. 2014. Effects of municipal solid waste compost on soil properties and vegetables growth. *Compost Science and Utilization* 22(3), 116–131. https://doi.org/10.1080 /1065657X.2014.899938.

Głąb, T., Żabiński, A., Sadowska, U., Gondek, K., Kopeć, M., Mierzwa-Hersztek, M., Tabor, S. and Stanek-Tarkowska, J. 2020. Fertilization effects of compost produced from maize, sewage sludge and biochar on soil water retention and chemical properties. *Soil and Tillage Research* 197, 104493. https://doi.org/10.1016/j.still .2019.104493.

Gómez-Brandón, M., Lazcano, C. and Domínguez, J. 2008. The evaluation of stability and maturity during the composting of cattle manure. *Chemosphere* 70(3), 436–444. https://doi.org/10.1016/j.chemosphere.2007.06.065.

Graham, R., Wortman, S. and Pittelkow, C. 2017. Comparison of organic and integrated nutrient management strategies for reducing soil N2O emissions. *Sustainability* 9(4), 510. https://doi.org/10.3390/su9040510.

Gravuer, K., Gennet, S. and Throop, H. L. 2019. Organic amendment additions to rangelands: a meta-analysis of multiple ecosystem outcomes. *Global Change Biology* 25(3), 1152–1170. https://doi.org/10.1111/gcb.14535.

Guenet, B., Gabrielle, B., Chenu, C., Arrouays, D., Balesdent, J., Bernoux, M., Bruni, E., Caliman, J., Cardinael, R., Chen, S., Ciais, P., Desbois, D., Fouche, J., Frank, S., Henault, C., Lugato, E., Naipal, V., Nesme, T., Obersteiner, M., Pellerin, S., Powlson, D. S., Rasse, D. P., Rees, F., Soussana, J., Su, Y., Tian, H., Valin, H. and Zhou, F. 2020. Can N2O emissions offset the benefits from soil organic carbon storage? *Global Change Biology* 27(2), 237–256. https://doi.org/10.1111/gcb.15342.

Habteweld, A. W., Brainard, D., Kravchenko, A., Grewal, P. S. and Melakeberhan, H. 2018. Effects of plant and animal waste-based compost amendments on the soil food web, soil properties, and yield and quality of fresh market and processing carrot cultivars. *Nematology* 20(2), 147–168. https://doi.org/10.1163/15685411-00003130.

Harada, Y. and Inoko, A. 1980. Relationship between cation-exchange capacity and degree of maturity of city refuse composts. *Soil Science and Plant Nutrition* 26(3), 353–362. https://doi.org/10.1080/00380768.1980.10431220.

Hargreaves, J., Adl, M. and Warman, P. 2008. A review of the use of composted municipal solid waste in agriculture. *Agriculture, Ecosystems and Environment* 123(1-3), 1–14. https://doi.org/10.1016/j.agee.2007.07.004.

Hartl, W., Putz, B. and Erhart, E. 2003. Influence of rates and timing of biowaste compost application on rye yield and soil nitrate levels. *European Journal of Soil Biology* 39(3), 129–139. https://doi.org/10.1016/S1164-5563(03)00028-1.

Hartz, T. K., Mitchell, J. P. and Giannini, C. 2000. Nitrogen and carbon mineralization dynamics of manures and composts. *HortScience* 35(2), 209–212. https://doi.org/10 .21273/HORTSCI.35.2.209.

Havlin, J. (Ed.) 2013. *Soil Fertility and Fertilizers: An Introduction to Nutrient Management* (8th ed.). Pearson, Upper Saddle River, NJ.

Haynes, R. J., Belyaeva, O. N. and Zhou, Y. F. 2015. Particle size fractionation as a method for characterizing the nutrient content of municipal green waste used for composting. *Waste Management* 35, 48–54.

Hazarika, J. and Khwairakpam, M. 2018. Evaluation of biodegradation feasibility through rotary drum composting recalcitrant primary paper mill sludge. *Waste Management* 76, 275–283. https://doi.org/10.1016/j.wasman.2018.03.044.

Helfenstein, J., Müller, I., Grüter, R., Bhullar, G., Mandloi, L., Papritz, A., Siegrist, M., Schulin, R. and Frossard, E. 2016. Organic wheat farming improves grain zinc concentration. *PLoS ONE* 11(8), e0160729. https://doi.org/10.1371/journal.pone.0160729.

Hemmat, A., Aghilinategh, N., Rezainejad, Y. and Sadeghi, M. 2010. Long-term impacts of municipal solid waste compost, sewage sludge and farmyard manure application on organic carbon, bulk density and consistency limits of a calcareous soil in central Iran. *Soil and Tillage Research* 108(1–2), 43–50. https://doi.org/10.1016/j.still.2010.03.007.

Horwath, W. R. and Paul, E. A. 1994. Microbial biomass. In: Weaver, R. W., Angle, S., Bottomley, P., Bezdicek, D., Smith, S., Tabatabai, A. and Wollum, A. (Eds), *SSSA Book Series*. Soil Science Society of America, Madison, WI, pp. 753–773. https://doi.org/10.2136/sssabookser5.2.c36.

Hurisso, T. T., Culman, S. W., Horwath, W. R., Wade, J., Cass, D., Beniston, J. W., Bowles, T. M., Grandy, A. S., Franzluebbers, A. J., Schipanski, M. E., Lucas, S. T. and Ugarte, C. M. 2016. Comparison of permanganate-oxidizable carbon and mineralizable carbon for assessment of organic matter stabilization and mineralization. *Soil Science Society of America Journal* 80(5), 1352–1364. https://doi.org/10.2136/sssaj2016.04.0106.

Hutchinson, G. L. and Mosier, A. R. 1981. Improved soil cover method for field measurement of nitrous oxide fluxes. *Soil Science Society of America Journal* 45(2), 311–316. https://doi.org/10.2136/sssaj1981.03615995004500020017x.

Iglesias-Jimenez, E. and Alvarez, C. E. 1993. Apparent availability of nitrogen in composted municipal refuse. *Biology and Fertility of Soils* 16(4), 313–318. https://doi.org/10.1007/BF00369312.

Insam, H. and de Bertoldi, M. 2007. Chapter 3 Microbiology of the composting process. In: *Waste Management Series*. Elsevier, pp. 25–48. https://doi.org/10.1016/S1478-7482(07)80006-6.

IPCC 2014. *2013 Supplement to the 2006 IPCC Guidelines for National Greenhouse Gas Inventories: Wetlands*. The Intergovernmental Panel on Climate Change (IPCC), Switzerland.

Jain, M. S., Daga, M. and Kalamdhad, A. S. 2018. Composting physics: a degradation process-determining tool for industrial sludge. *Ecological Engineering* 116, 14–20. https://doi.org/10.1016/j.ecoleng.2018.02.015.

Jeger, M. J. and Viljanen-Rollinson, S. L. H. 2001. The use of the area under the disease-progress curve (AUDPC) to assess quantitative disease resistance in crop cultivars. *Theoretical and Applied Genetics* 102(1), 32–40. https://doi.org/10.1007/s001220051615.

Jindo, K., Chocano, C., Melgares de Aguilar, J., González, D., Hernandez, T. and García, C. 2016. Impact of compost application during 5 years on crop production, soil microbial activity, carbon fraction, and humification process. *Communications in Soil Science and Plant Analysis*. https://doi.org/10.1080/00103624.2016.1206922.

Kanaan, H., Hadar, Y., Medina, S., Krasnovsky, A., Mordechai-Lebiush, S., Tsror, L., Katan, J. and Raviv, M. 2018. Effect of compost properties on progress rate of Verticillium dahliae attack on eggplant (Solanum melongena L.). Compost Science and Utilization 26(2), 71–78. https://doi.org/10.1080/1065657X.2017.1366375.

Karlen, D. L., Veum, K. S., Sudduth, K. A., Obrycki, J. F. and Nunes, M. R. 2019. Soil health assessment: past accomplishments, current activities, and future opportunities. Soil and Tillage Research 195, 104365. https://doi.org/10.1016/j.still.2019.104365.

Kaza, S., Yao, L., Bhada-Tata, P. and Van Woerden, F. 2018. What a Waste 2.0: A Global Snapshot of Solid Waste Management to 2050. The World Bank. https://doi.org/10.1596/978-1-4648-1329-0.

Kibblewhite, M. G., Ritz, K. and Swift, M. J. 2008. Soil health in agricultural systems. Philosophical Transactions of the Royal Society of London. Series B, Biological Sciences 363(1492), 685–701. https://doi.org/10.1098/rstb.2007.2178.

Kim, B. R., Shin, J., Guevarra, R. B., Lee, J. H., Kim, D. W., Seol, K. H., Lee, J. H., Kim, H. B. and Isaacson, R. E. 2017. Deciphering diversity indices for a better understanding of microbial communities. Journal of Microbiology and Biotechnology 27(12), 2089–2093. https://doi.org/10.4014/jmb.1709.09027.

Kim, H.-Y., Lim, S.-S., Kwak, J.-H., Lee, S.-I., Lee, D.-S., Hao, X., Yoon, K.-S. and Choi, W.-J. 2011. Soil and compost type affect phosphorus leaching from inceptisol, ultisol, and andisol in a column experiment. Communications in Soil Science and Plant Analysis 42(18), 2188–2199. https://doi.org/10.1080/00103624.2011.602450.

Knapp, B. A., Ros, M. and Insam, H. 2010. Do composts affect the soil microbial community? In: Insam, H., Franke-Whittle, I. and Goberna, M. (Eds), Microbes at Work: From Wastes to Resources. Springer, Berlin, Heidelberg, pp. 271–291. https://doi.org/10.1007/978-3-642-04043-6_14.

Koike, S. T. 2008. Crown rot of strawberry caused by Macrophomina phaseolina in California. Plant Disease 92(8), 1253. https://doi.org/10.1094/PDIS-92-8-1253B.

Kranz, C. N., McLaughlin, R. A., Johnson, A., Miller, G. and Heitman, J. L. 2020. The effects of compost incorporation on soil physical properties in urban soils – a concise review. Journal of Environmental Management 261, 110209. https://doi.org/10.1016/j.jenvman.2020.110209.

Krause, H.-M., Fliessbach, A., Mayer, J. and Mäder, P. 2020. Implementation and management of the DOK long-term system comparison trial. In: Long-Term Farming Systems Research. Elsevier, pp. 37–51. https://doi.org/10.1016/B978-0-12-818186-7.00003-5.

Kumar, S. 2011. Composting of municipal solid waste. Critical Reviews in Biotechnology 31(2), 112–136. https://doi.org/10.3109/07388551.2010.492207.

Lairon, D. 2011. Nutritional quality and safety of organic food. In: Lichtfouse, E., Hamelin, M., Navarrete, M. and Debaeke, P. (Eds), Sustainable Agriculture (vol. 2). Springer Netherlands, Dordrecht, pp. 99–110. https://doi.org/10.1007/978-94-007-0394-0_7.

Lal, R. 2004. Soil carbon sequestration to mitigate climate change. Geoderma 123(1–2), 1–22. https://doi.org/10.1016/j.geoderma.2004.01.032.

Langdon, K. A., Chandra, A., Bowles, K., Symons, A., Pablo, F. and Osborne, K. 2019. A preliminary ecological and human health risk assessment for organic contaminants in composted municipal solid waste generated in New South Wales, Australia. Waste Management 100, 199–207. https://doi.org/10.1016/j.wasman.2019.09.001.

Lannan, A. P., Erich, M. S. and Ohno, T. 2013. Compost feedstock and maturity level affect soil response to amendment. *Biology and Fertility of Soils* 49(3), 273–285. https://doi .org/10.1007/s00374-012-0715-0.

Latifah, O., Ahmed, O. H. and Majid, N. M. A. 2018. Soil pH buffering capacity and nitrogen availability following compost application in a tropical acid soil. *Compost Science and Utilization* 26(1), 1–15. https://doi.org/10.1080/1065657X.2017 .1329039.

Lazcano, C., Zhu-Barker, X. and Decock, C. 2021a. Effects of organic fertilizers on the soil microorganisms responsible for N2O emissions: a review. *Microorganisms* 9(5), 983. https://doi.org/10.3390/microorganisms9050983.

Lazcano, C., Boyd, E., Holmes, G., Hewavitharana, S., Pasulka, A. and Ivors, K. 2021b. The rhizosphere microbiome plays a role in the resistance to soil-borne pathogens and nutrient uptake of strawberry cultivars under field conditions. *Scientific Reports* 11(1), 3188. https://doi.org/10.1038/s41598-021-82768-2.

Lazcano, C., Gómez-Brandón, M., Revilla, P. and Domínguez, J. 2013. Short-term effects of organic and inorganic fertilizers on soil microbial community structure and function: A field study with sweet corn. *Biology and Fertility of Soils* 49(6), 723–733. https://doi .org/10.1007/s00374-012-0761-7.

Lazicki, P., Geisseler, D. and Lloyd, M. 2020. Nitrogen mineralization from organic amendments is variable but predictable. *Journal of Environmental Quality* 49(2), 483–495. https://doi.org/10.1002/jeq2.20030.

Leclerc, B., Georges, P., Cauwel, B. and Lairon, D. 1995. A five year study on nitrate leaching under crops fertilised with mineral and organic fertilisers in lysimeters. *Biological Agriculture and Horticulture* 11(1–4), 301–308. https://doi.org/10.1080 /01448765.1995.9754714.

Leelamanie, D. A. L. and Manawardana, C. U. 2019. Soil hydrophysical properties as affected by solid waste compost amendments: seasonal and short-term effects in an ultisol. *Journal of Hydrology and Hydromechanics* 67(3), 232–239. https://doi.org /10.2478/johh-2019-0007.

Lehman, R. M., Acosta-Martinez, V., Buyer, J. S., Cambardella, C. A., Collins, H. P., Ducey, T. F., Halvorson, J. J., Jin, V. L., Johnson, J. M. F., Kremer, R. J., Lundgren, J. G., Manter, D. K., Maul, J. E., Smith, J. L. and Stott, D. E. 2015. Soil biology for resilient, healthy soil. *Journal of Soil and Water Conservation* 70(1), 12A–18A. https://doi.org/10.2489 /jswc.70.1.12A.

Lehmann, J. and Kleber, M. 2015. The contentious nature of soil organic matter. *Nature* 528(7580), 60–68. https://doi.org/10.1038/nature16069.

Lehtinen, T., Dersch, G., Söllinger, J., Baumgarten, A., Schlatter, N., Aichberger, K. and Spiegel, H. 2017. Long-term amendment of four different compost types on a loamy silt cambisol: impact on soil organic matter, nutrients and yields. *Archives of Agronomy and Soil Science* 63(5), 663–673. https://doi.org/10.1080/03650340 .2016.1235264.

Leroy, B. L. M., Herath, M. S. K., De Neve, S., Gabriels, D., Bommele, L., Reheul, D. and Moens, M. 2008. Effect of vegetable, fruit and garden (VFG) waste compost on soil physical properties. *Compost Science and Utilization* 16(1), 43–51. https://doi.org /10.1080/1065657X.2008.10702354.

Li, H., Van den Bulcke, J., Wang, X., Gebremikael, M. T., Hagan, J., De Neve, S. and Sleutel, S. 2020. Soil texture strongly controls exogenous organic matter mineralization indirectly via moisture upon progressive drying – evidence from incubation experiments. *Soil Biology and Biochemistry* 151, 108051. https://doi.org/10.1016/j .soilbio.2020.108051.

Li, J., Cooper, J. M., Lin, Z., Li, Y., Yang, X. and Zhao, B. 2015. Soil microbial community structure and function are significantly affected by long-term organic and mineral fertilization regimes in the North China Plain. *Applied Soil Ecology* 96, 75–87. https:// doi.org/10.1016/j.apsoil.2015.07.001.

Li, P., Lang, M., Li, C. and Hao, X. 2016. Nitrous oxide and carbon dioxide emissions from soils amended with compost and manure from cattle fed diets containing wheat dried distillers' grains with solubles. *Canadian Journal of Soil Science*, CJSS-2016-0068. https://doi.org/10.1139/CJSS-2016-0068.

Li, Y. C., Stoffella, P. J., Alva, A. K., Calvert, D. V. and Graetz, D. A. 1997. Leaching of nitrate, ammonium, and phosphate From compost amended soil columns. *Compost Science and Utilization* 5(2), 63–67. https://doi.org/10.1080/1065657X.1997.10701875.

Lindahl, B. D., Nilsson, R. H., Tedersoo, L., Abarenkov, K., Carlsen, T., Kjøller, R., Kõljalg, U., Pennanen, T., Rosendahl, S., Stenlid, J. and Kauserud, H. 2013. Fungal community analysis by high-throughput sequencing of amplified markers – a user's guide. *New Phytologist* 199(1), 288–299. https://doi.org/10.1111/nph.12243.

Liu, H., Wang, Z. H., Li, F., Li, K., Yang, N., Yang, Y., Huang, D., Liang, D., Zhao, H., Mao, H., Liu, J. and Qiu, W. 2014. Grain iron and zinc concentrations of wheat and their relationships to yield in major wheat production areas in China. *Field Crops Research* 156, 151–160. https://doi.org/10.1016/j.fcr.2013.11.011.

Liu, J., Calderón, F. J. and Fonte, S. J. 2021. Compost inputs, cropping system, and rotation phase drive aggregate-associated carbon. *Soil Science Society of America Journal* 85(3), 829–846. https://doi.org/10.1002/saj2.20252.

Liu, L., Wang, S., Guo, X. and Wang, H. 2019. Comparison of the effects of different maturity composts on soil nutrient, plant growth and heavy metal mobility in the contaminated soil. *Journal of Environmental Management* 250, 109525. https://doi .org/10.1016/j.jenvman.2019.109525.

Liu, Q., Wang, J., Bai, Z., Ma, L. and Oenema, O. 2017. Global animal production and nitrogen and phosphorus flows. *Soil Research* 55(6), 451. https://doi.org/10.1071 /SR17031.

Liu, T., Chen, X., Hu, F., Ran, W., Shen, Q., Li, H. and Whalen, J. K. 2016. Carbon-rich organic fertilizers to increase soil biodiversity: evidence from a meta-analysis of nematode communities. *Agriculture, Ecosystems and Environment* 232, 199–207. https://doi .org/10.1016/j.agee.2016.07.015.

Liu, Z., Wang, X., Wang, F., Bai, Z., Chadwick, D., Misselbrook, T. and Ma, L. 2020. The progress of composting technologies from static heap to intelligent reactor: benefits and limitations. *Journal of Cleaner Production* 270, 122328. https://doi.org/10.1016 /j.jclepro.2020.122328.

Lloyd, M. 2016. Growing for the future: collective action, land stewardship and soilborne pathogens in California strawberry production. *California Agriculture* 70(3), 101–103. https://doi.org/10.3733/ca.2016a0009.

Lloyd, M., Kluepfel, D. and Gordon, T. 2016. Evaluation of four commercial composts on strawberry plant productivity and soil characteristics in California. *International Journal of Fruit Science* 16(sup1), 84–107. https://doi.org/10.1080/15538362.2016 .1239562.

Lodha, S., Sharma, S. K. and Aggarwal, R. K. 2002. Inactivation of Macrophomina phaseolina propagules during composting and effect of composts on dry root rot severity and on seed yield of clusterbean. *European Journal of Plant Pathology* 108(3), 253–261. https://doi.org/10.1023/A:1015103315068.

Logsdon, S. D., Sauer, P. A. and Shipitalo, M. J. 2017. Compost improves urban soil and water quality. *Journal of Water Resource and Protection* 09(4), 345–357. https://doi .org/10.4236/jwarp.2017.94023.

Lü, H., Chen, X. H., Mo, C. H., Huang, Y. H., He, M. Y., Li, Y. W., Feng, N. X., Katsoyiannis, A. and Cai, Q. Y. 2021. Occurrence and dissipation mechanism of organic pollutants during the composting of sewage sludge: a critical review. *Bioresource Technology* 328, 124847. https://doi.org/10.1016/j.biortech.2021.124847.

Lucas, J. A., Hawkins, N. J. and Fraaije, B. A. 2015. The evolution of fungicide resistance. In: *Advances in Applied Microbiology*. Elsevier, pp. 29–92. https://doi.org/10.1016/ bs.aambs.2014.09.001.

Lugato, E., Leip, A. and Jones, A. 2018. Mitigation potential of soil carbon management overestimated by neglecting N2O emissions. *Nature Climate Change* 8(3), 219–223. https://doi.org/10.1038/s41558-018-0087-z.

Madden, L. V., Hughes, G. and van den Bosch, F. 2017. *The Study of Plant Disease Epidemics*. The American Phytopathological Society. https://doi.org/10.1094 /9780890545058.

Makan, A. and Fadili, A. 2020. Sustainability assessment of large-scale composting technologies using PROMETHEE method. *Journal of Cleaner Production* 261, 121244. https://doi.org/10.1016/j.jclepro.2020.121244.

Maltais-Landry, G., Scow, K., Brennan, E., Torbert, E. and Vitousek, P. 2016. Higher flexibility in input N:P ratios results in more balanced phosphorus budgets in two long-term experimental agroecosystems. *Agriculture, Ecosystems and Environment* 223, 197–210. https://doi.org/10.1016/j.agee.2016.03.007.

Maltais-Landry, G., Scow, K., Brennan, E. and Vitousek, P. 2015. Long-term effects of compost and cover crops on soil phosphorus in two California agroecosystems. *Soil Science Society of America Journal* 79(2), 688–697. https://doi.org/10.2136/ sssaj2014.09.0369.

Mamo, M., Moncrief, J. F., Rosen, C. J. and Halbach, T. R. 2000. The effect of municipal solid waste compost application on soil water and water stress in irrigated corn. *Compost Science and Utilization* 8(3), 236–246. https://doi.org/10.1080/1065657X .2000.10701996.

Mamo, M., Rosen, C. J. and Halbach, T. R. 1999. Nitrogen availability and leaching from soil amended with municipal solid waste compost. *Journal of Environmental Quality* 28(4), 1074–1082. https://doi.org/10.2134/jeq1999.00472425002800040003x.

Mangalassery, S., Kalaivanan, D. and Philip, P. S. 2019. Effect of inorganic fertilisers and organic amendments on soil aggregation and biochemical characteristics in a weathered tropical soil. *Soil and Tillage Research* 187, 144–151. https://doi.org/10 .1016/j.still.2018.12.008.

Martínez-Blanco, J., Antón, A., Rieradevall, J., Castellari, M. and Muñoz, P. 2011. Comparing nutritional value and yield as functional units in the environmental assessment of horticultural production with organic or mineral fertilization: the case of Mediterranean cauliflower production. *International Journal of Life Cycle Assessment* 16(1), 12–26. https://doi.org/10.1007/s11367-010-0238-6.

Martínez-Blanco, J., Lazcano, C., Boldrin, A., Muñoz, P., Rieradevall, J., Møller, J., Antón, A. and Christensen, T. H. 2013a. Assessing the environmental benefits of compost use-on-land through an LCA perspective. In: Lichtfouse, E. (Ed.), *Sustainable Agriculture*

Reviews, Sustainable Agriculture Reviews. Springer Netherlands, Dordrecht, pp. 255–318. https://doi.org/10.1007/978-94-007-5961-9_9.

Martínez-Blanco, J., Lazcano, C., Christensen, T. H., Muñoz, P., Rieradevall, J., Møller, J., Antón, A. and Boldrin, A. 2013b. Compost benefits for agriculture evaluated by life cycle assessment. A review. Agronomy for Sustainable Development 33(4), 721–732. https://doi.org/10.1007/s13593-013-0148-7.

McCarthy, G., Lawlor, P. G., Coffey, L., Nolan, T., Gutierrez, M. and Gardiner, G. E. 2011. An assessment of pathogen removal during composting of the separated solid fraction of pig manure. Bioresource Technology 102(19), 9059–9067. https://doi.org/10.1016/j.biortech.2011.07.021.

McBride, M. B. 1995. Toxic metal accumulation from agricultural use of sludge: are USEPA regulations protective? Journal of Environmental Quality 24(1), 5–18. https://doi.org/10.2134/jeq1995.00472425002400010002x.

McDowell, R. W. and Sharpley, A. N. 2004. Variation of phosphorus leached from Pennsylvanian soils amended with manures, composts or inorganic fertilizer. Agriculture, Ecosystems and Environment 102(1), 17–27. https://doi.org/10.1016/j.agee.2003.07.002.

Mehta, C. M., Palni, U., Franke-Whittle, I. H. and Sharma, A. K. 2014. Compost: its role, mechanism and impact on reducing soil-borne plant diseases. Waste Management 34(3), 607–622. https://doi.org/10.1016/j.wasman.2013.11.012.

Meng, L., Ding, W. and Cai, Z. 2005. Long-term application of organic manure and nitrogen fertilizer on N2O emissions, soil quality and crop production in a sandy loam soil. Soil Biology and Biochemistry 37(11), 2037–2045. https://doi.org/10.1016/j.soilbio.2005.03.007.

Miller, J. J., Beasley, B. W., Drury, C. F., Larney, F. J. and Hao, X. 2015. Influence of long-term (9 yr) composted and stockpiled feedlot manure application on selected soil physical properties of a clay loam soil in Southern Alberta. Compost Science and Utilization 23(1), 1–10. https://doi.org/10.1080/1065657X.2014.963741.

Minasny, B. and McBratney, A. B. 2018. Limited effect of organic matter on soil available water capacity. European Journal of Soil Science 69(1), 39–47. https://doi.org/10.1111/ejss.12475.

Moebius-Clune, B. N. 2016. Comprehensive assessment of soil health: the Cornell framework manual. Cornell University, Ithaca, New York. Available at: http://www.css.cornell.edu/extension/soil-health/manual.pdf

Mohammadshirazi, F., McLaughlin, R. A., Heitman, J. L. and Brown, V. K. 2017. A multi-year study of tillage and amendment effects on compacted soils. Journal of Environmental Management 203(1), 533–541. https://doi.org/10.1016/j.jenvman.2017.07.031.

Morlat, R. and Symoneaux, R. 2008. Long-term additions of organic amendments in a Loire Valley vineyard on a calcareous sandy soil. III. Effects on fruit composition and chemical and sensory characteristics of Cabernet Franc wine. American Journal of Enology and Viticulture 59, 375–386.

Mugnai, S., Masi, E., Azzarello, E. and Mancuso, S. 2012. Influence of long-term application of green waste compost on soil characteristics and growth, yield and quality of grape (Vitis vinifera L.). Compost Science and Utilization 20(1), 29–33. https://doi.org/10.1080/1065657X.2012.10737019.

Mylavarapu, R. S. and Zinati, G. M. 2009. Improvement of soil properties using compost for optimum parsley production in sandy soils. Scientia Horticulturae 120(3), 426–430. https://doi.org/10.1016/j.scienta.2008.11.038.

Nair, A. and Ngouajio, M. 2012. Soil microbial biomass, functional microbial diversity, and nematode community structure as affected by cover crops and compost in an organic vegetable production system. *Applied Soil Ecology* 58, 45–55. https://doi.org/10.1016/j.apsoil.2012.03.008.

Nannipieri, P., Ceccanti, B., Cervelli, S. and Matarese, E. 1980. Extraction of phosphatase, urease, proteases, organic carbon, and nitrogen from soil. *Soil Science Society of America Journal* 44(5), 1011–1016. https://doi.org/10.2136/sssaj1980.03615995004400050028x.

Neufeld, J. D., Dumont, M. G., Vohra, J. and Murrell, J. C. 2007. Methodological considerations for the use of stable isotope probing in microbial ecology. *Microbial Ecology* 53(3), 435–442. https://doi.org/10.1007/s00248-006-9125-x.

Noble, R. and Coventry, E. 2005. Suppression of soil-borne plant diseases with composts: a review. *Biocontrol Science and Technology* 15(1), 3–20. https://doi.org/10.1080/09583150400015904.

Norris, C. E., Bean, G. M., Cappellazzi, S. B., Cope, M., Greub, K. L. H., Liptzin, D., Rieke, E. L., Tracy, P. W., Morgan, C. L. S. and Honeycutt, C. W. 2020. Introducing the North American project to evaluate soil health measurements. *Agronomy Journal* 112, 3195–3215. https://doi.org/10.1002/agj2.20234.

Nunes, M. R., Karlen, D. L., Veum, K. S., Moorman, T. B. and Cambardella, C. A. 2020. Biological soil health indicators respond to tillage intensity: a US meta-analysis. *Geoderma* 369, 114335. https://doi.org/10.1016/j.geoderma.2020.114335.

Oliveira, M. and Duarte, E. 2016. Integrated approach to winery waste: waste generation and data consolidation. *Frontiers of Environmental Science and Engineering* 10(1), 168–176. https://doi.org/10.1007/s11783-014-0693-6.

Oudart, D., Paul, E., Robin, P. and Paillat, J. M. 2012. Modeling organic matter stabilization during windrow composting of livestock effluents. *Environmental Technology* 33(19–21), 2235–2243. https://doi.org/10.1080/09593330.2012.728736.

Ouédraogo, E. 2001. Use of compost to improve soil properties and crop productivity under low input agricultural system in West Africa. *Agriculture, Ecosystems and Environment* 84(3), 259–266. https://doi.org/10.1016/S0167-8809(00)00246-2.

Pai, S., Ai, N. and Zheng, J. 2019. Decentralized community composting feasibility analysis for residential food waste: a Chicago case study. *Sustainable Cities and Society* 50, 101683. https://doi.org/10.1016/j.scs.2019.101683.

Parkin, T. B. and Venterea, R. T. 2010. Chapter 3. Chamber-based trace gas flux measurements. In: R. F. Follett (Ed.), Sampling Protocols. USDA-ARS, pp. 1–39. Available at: (https://www.ars.usda.gov/natural-resources-and-sustainable-agricultural-systems/soil-and-air/docs/gracenet-sampling-protocols/

Paustian, K., Larson, E., Kent, J., Marx, E. and Swan, A. 2019. Soil C sequestration as a biological negative emission strategy. *Frontiers in Climate* 1, 8. https://doi.org/10.3389/fclim.2019.00008.

Pawlett, M., Hannam, J. A. and Knox, J. W. 2021. Redefining soil health. *Microbiology* 167(1). https://doi.org/10.1099/mic.0.001030.

Payen, F. T., Sykes, A., Aitkenhead, M., Alexander, P., Moran, D. and MacLeod, M. 2021. Predicting the abatement rates of soil organic carbon sequestration management in Western European vineyards using random forest regression. *Cleaner Environmental Systems* 2, 100024. https://doi.org/10.1016/j.cesys.2021.100024.

Pérez-Piqueres, A., Edel-Hermann, V., Alabouvette, C. and Steinberg, C. 2006. Response of soil microbial communities to compost amendments. *Soil Biology and Biochemistry* 38(3), 460–470. https://doi.org/10.1016/j.soilbio.2005.05.025.

Pergola, M., Persiani, A., Palese, A. M., Di Meo, V., Pastore, V., D'Adamo, C. and Celano, G. 2018. Composting: the way for a sustainable agriculture. *Applied Soil Ecology* 123, 744–750. https://doi.org/10.1016/j.apsoil.2017.10.016.

Petrie, B., Barden, R. and Kasprzyk-Hordern, B. 2015. A review on emerging contaminants in wastewaters and the environment: current knowledge, understudied areas and recommendations for future monitoring. *Water Research* 72, 3–27. https://doi.org/10.1016/j.watres.2014.08.053.

Pezzolla, D., Said-Pullicino, D., Raggi, L., Albertini, E. and Gigliotti, G. 2013. Short-term variations in labile organic C and microbial biomass activity and structure After organic amendment of arable soils. *Soil Science* 178(9), 474–485. https://doi.org/10.1097/SS.0000000000000012.

Pugliese, M., Gilardi, G., Garibaldi, A. and Gullino, M. L. 2015. Organic amendments and soil suppressiveness: results with vegetable and ornamental crops. In: Meghvansi, M. K. and Varma, A. (Eds), *Organic Amendments and Soil Suppressiveness in Plant Disease Management, Soil Biology*. Springer International Publishing, Cham, pp. 495–509. https://doi.org/10.1007/978-3-319-23075-7_24.

Ramos, M. C. 2017. Effects of compost amendment on the available soil water and grape yield in vineyards planted after land levelling. *Agricultural Water Management* 191, 67–76. https://doi.org/10.1016/j.agwat.2017.05.013.

Ramos, T. M., Jay-Russell, M. T., Millner, P. D., Shade, J., Misiewicz, T., Sorge, U. S., Hutchinson, M., Lilley, J. and Pires, A. F. A. 2019. Assessment of biological soil amendments of animal origin use, research needs, and extension opportunities in organic production. *Frontiers in Sustainable Food Systems* 3, 73. https://doi.org/10.3389/fsufs.2019.00073.

Ramzani, P. M. A., Shan, L., Anjum, S., Khan, W. U., Ronggui, H., Iqbal, M., Virk, Z. A. and Kausar, S. 2017. Improved quinoa growth, physiological response, and seed nutritional quality in three soils having different stresses by the application of acidified biochar and compost. *Plant Physiology and Biochemistry* 116, 127–138. https://doi.org/10.1016/j.plaphy.2017.05.003.

Rawls, W. J., Pachepsky, Y. A., Ritchie, J. C., Sobecki, T. M. and Bloodworth, H. 2003. Effect of soil organic carbon on soil water retention. *Geoderma* 116(1-2), 61–76. https://doi.org/10.1016/S0016-7061(03)00094-6.

Rinot, O., Levy, G. J., Steinberger, Y., Svoray, T. and Eshel, G. 2019. Soil health assessment: a critical review of current methodologies and a proposed new approach. *Science of the Total Environment* 648, 1484–1491. https://doi.org/10.1016/j.scitotenv.2018.08.259.

Rivers, E. N., Heitman, J. L., McLaughlin, R. A. and Howard, A. M. 2021. Reducing roadside runoff: tillage and compost improve stormwater mitigation in urban soils. *Journal of Environmental Management* 280, 111732. https://doi.org/10.1016/j.jenvman.2020.111732.

Ros, M., Hernandez, M. T. and Garcia, C. 2003. Soil microbial activity after restoration of a semiarid soil by organic amendments. *Soil Biology and Biochemistry* 35(3), 463–469. https://doi.org/10.1016/S0038-0717(02)00298-5.

Ros, M., Klammer, S., Knapp, B., Aichberger, K. and Insam, H. 2006. Long-term effects of compost amendment of soil on functional and structural diversity and microbial activity. *Soil Use and Management* 22(2), 209–218. https://doi.org/10.1111/j.1475 -2743.2006.00027.x.

Rubio, R., Pérez-Murcia, M. D., Agulló, E., Bustamante, M. A., Sánchez, C., Paredes, C. and Moral, R. 2013. Recycling of agro-food wastes into vineyards by composting: agronomic validation in field conditions. *Communications in Soil Science and Plant Analysis* 44(1–4), 502–516. https://doi.org/10.1080/00103624.2013.744152.

Ruehlmann, J. and Körschens, M. 2009. Calculating the effect of soil organic matter concentration on soil bulk density. *Soil Science Society of America Journal* 73(3), 876–885. https://doi.org/10.2136/sssaj2007.0149.

Ruggieri, L., Cadena, E., Martínez-Blanco, J., Gasol, C. M., Rieradevall, J., Gabarrell, X., Gea, T., Sort, X. and Sánchez, A. 2009. Recovery of organic wastes in the Spanish wine industry. Technical, economic and environmental analyses of the composting process. *Journal of Cleaner Production* 17(9), 830–838. https://doi.org/10.1016/J .jclepro.2008.12.005.

Ryals, R., Kaiser, M., Torn, M. S., Berhe, A. A. and Silver, W. L. 2014. Impacts of organic matter amendments on carbon and nitrogen dynamics in grassland soils. *Soil Biology and Biochemistry* 68, 52–61. https://doi.org/10.1016/j.soilbio.2013.09 .011.

Saha, S., Pandey, A. K., Gopinath, K. A., Bhattacharaya, R., Kundu, S. and Gupta, H. S. 2007. Nutritional quality of organic rice grown on organic composts. *Agronomy for Sustainable Development* 27(3), 223–229. https://doi.org/10.1051/agro:2007002.

Saison, C., Degrange, V., Oliver, R., Millard, P., Commeaux, C., Montange, D. and Le Roux, X. 2006. Alteration and resilience of the soil microbial community following compost amendment: effects of compost level and compost-borne microbial community. *Environmental Microbiology* 8(2), 247–257. https://doi.org/10.1111/j.1462-2920 .2005.00892.x.

Salomé, C., Coll, P., Lardo, E., Metay, A., Villenave, C., Marsden, C., Blanchart, E., Hinsinger, P. and Le Cadre, E. 2016. The soil quality concept as a framework to assess management practices in vulnerable agroecosystems: a case study in Mediterranean vineyards. *Ecological Indicators* 61, 456–465. https://doi.org/10 .1016/j.ecolind.2015.09.047.

Sax, M. S., Bassuk, N., van Es, H. and Rakow, D. 2017. Long-term remediation of compacted urban soils by physical fracturing and incorporation of compost. *Urban Forestry and Urban Greening* 24, 149–156. https://doi.org/10.1016/j.ufug.2017.03.023.

Schimel, J. P. and Bennett, J. 2004. Nitrogen mineralization: challenges of a changing paradigm. *Ecology* 85(3), 591–602. https://doi.org/10.1890/03-8002.

Şeker, C. and Manirakiza, N. 2020. Effectiveness of compost and biochar in improving water retention characteristics and aggregation of a sandy clay loam soil under wind erosion. *Carpathian Journal of Earth and Environmental Sciences* 15(1), 5–18. https:// doi.org/10.26471/cjees/2020/015/103.

Sesay, A. A., Lasaridi, K., Stentiford, E. and Budd, T. 1997. Controlled composting of paper pulp sludge using the aerated static pile method. *Compost Science and Utilization* 5(1), 82–96. https://doi.org/10.1080/1065657X.1997.10701866.

Seufert, V. and Ramankutty, N. 2017. Many shades of gray–the context-dependent performance of organic agriculture. *Science Advances* 3(3), e1602638. https://doi .org/10.1126/sciadv.1602638.

Sharma, S., Aneja, M. K., Mayer, J., Munch, J. C. and Schloter, M. 2005. Diversity of transcripts of nitrite reductase genes (*nirK* and *nirS*) in rhizospheres of grain legumes.

Applied and Environmental Microbiology 71(4), 2001–2007. https://doi.org/10.1128/AEM.71.4.2001-2007.2005.

Shaw, D. V., Gordon, T. R., Hansen, J. and Kirkpatrick, S. C. 2010. Relationship between the extent of colonization by *Verticillium dahliae* and symptom expression in strawberry (*Fragaria × ananassa*) genotypes resistant to Verticillium wilt. *Plant Pathology* 59(2), 376–381. https://doi.org/10.1111/j.1365-3059.2009.02203.x.

Shaw, D. V., Gubler, W. D. and Hansen, J. 1997. Field resistance of California strawberries to Verticilium dahliae at three conidial innoculum concentrations. *HortScience* 32(4), 711–713.

Shiralipour, A., McConnell, D. B. and Smith, W. H. 1992. Physical and chemical properties of soils as affected by municipal solid waste compost application. *Biomass and Bioenergy* 3(3–4), 261–266. https://doi.org/10.1016/0961-9534(92)90030-T.

Siedt, M., Schäffer, A., Smith, K. E. C., Nabel, M., Roß-Nickoll, M. and van Dongen, J. T. 2021. Comparing straw, compost, and biochar regarding their suitability as agricultural soil amendments to affect soil structure, nutrient leaching, microbial communities, and the fate of pesticides. *Science of the Total Environment* 751, 141607. https://doi.org/10.1016/j.scitotenv.2020.141607.

Siles-Castellano, A. B., López, M. J., López-González, J. A., Suárez-Estrella, F., Jurado, M. M., Estrella-González, M. J. and Moreno, J. 2020. Comparative analysis of phytotoxicity and compost quality in industrial composting facilities processing different organic wastes. *Journal of Cleaner Production* 252, 119820. https://doi.org/10.1016/j.jclepro.2019.119820.

Six, J., Bossuyt, H., Degryze, S. and Denef, K. 2004. A history of research on the link between (micro)aggregates, soil biota, and soil organic matter dynamics. *Soil and Tillage Research* 79(1), 7–31. https://doi.org/10.1016/j.still.2004.03.008.

Six, J., Elliott, E. T. and Paustian, K. 2000. Soil macroaggregate turnover and microaggregate formation: a mechanism for C sequestration under no-tillage agriculture. *Soil Biology and Biochemistry* 32(14), 2099–2103. https://doi.org/10.1016/S0038-0717(00)00179-6.

Solaiman, Z. M., Yang, H., Archdeacon, D., Tippett, O., Tibi, M. and Whiteley, A. S. 2019. Humus-rich compost increases lettuce growth, nutrient uptake, mycorrhizal colonisation, and soil fertility. *Pedosphere* 29(2), 170–179. https://doi.org/10.1016/S1002-0160(19)60794-0.

Solomon, S., Intergovernmental Panel on Climate Change and Intergovernmental Panel on Climate Change (Eds) 2007. Climate Change *2007: The Physical Science Basis: Contribution of Working Group I to the Fourth Assessment Report of the Intergovernmental Panel on Climate Change*. Cambridge University Press, Cambridge. New York.

Stamatiadis, S., Werner, M. and Buchanan, M. 1999. Field assessment of soil quality as affected by compost and fertilizer application in a broccoli field (San Benito County, California). *Applied Soil Ecology* 12(3), 217–225. https://doi.org/10.1016/S0929-1393(99)00013-X.

Stoffella, P., He, Z., Kahn, B., Yang, X. and Calvert, D. 2001. Plant nutrition benefits of phosphorus, potassium, calcium, magnesium, and micronutrients from compost utilization. In: Stoffella, P. and Kahn, B. (Eds), *Compost Utilization in Horticultural Cropping Systems*. CRC Press. https://doi.org/10.1201/9781420026221.ch15.

Tabatabai, M. A. 1994. Soil enzymes. In: Weaver, R. W., Angle, S., Bottomley, P., Bezdicek, D., Smith, S., Tabatabai, A. and Wollum, A. (Eds), *SSSA Book Series*. Soil Science Society of America, Madison, WI, pp. 775-833. https://doi.org/10.2136/sssabookser5.2.c37.

Tautges, N. and Scow, K. 2020. Pursuing agroecosystem resilience in a long-term Mediterranean agricultural experiment. In: *Long-Term Farming Systems Research*. Elsevier, pp. 53-69. https://doi.org/10.1016/B978-0-12-818186-7.00004-7.

Tejada, M. and Gonzalez, J. L. 2006a. The relationships between erodibility and erosion in a soil treated with two organic amendments. *Soil and Tillage Research* 91(1-2), 186-198. https://doi.org/10.1016/j.still.2005.12.003.

Tejada, M. and Gonzalez, J. L. 2006b. Crushed cotton gin compost on soil biological properties and rice yield. *European Journal of Agronomy* 25(1), 22-29. https://doi.org/10.1016/j.eja.2006.01.007.

Tejada, M., Gonzalez, J. L., García-Martínez, A. M. and Parrado, J. 2008. Application of a green manure and green manure composted with beet vinasse on soil restoration: effects on soil properties. *Bioresource Technology* 99(11), 4949-4957. https://doi.org/10.1016/j.biortech.2007.09.026.

Temiz, C., Akca, M. O., Cayci, G. and Baran, A. 2021. Assessment of the effect of different compost materials on aggregation and mechanical properties in an entisol. *Communications in Soil Science and Plant Analysis* 52(8), 871-885. https://doi.org/10.1080/00103624.2020.1869770.

Tian, W., Wang, L., Li, Y., Zhuang, K., Li, G., Zhang, J., Xiao, X. and Xi, Y. 2015. Responses of microbial activity, abundance, and community in wheat soil after three years of heavy fertilization with manure-based compost and inorganic nitrogen. *Agriculture, Ecosystems and Environment* 213, 219-227. https://doi.org/10.1016/j.agee.2015.08.009.

Tiefenbacher, A., Sandén, T., Haslmayr, H.-P., Miloczki, J., Wenzel, W. and Spiegel, H. 2021. Optimizing carbon sequestration in croplands: a synthesis. *Agronomy* 11(5), 882. https://doi.org/10.3390/agronomy11050882.

Torres, I. F., Bastida, F., Hernández, T., Albaladejo, J. and García, C. 2015. Enzyme activity, microbial biomass and community structure in a long-term restored soil under semi-arid conditions. *Soil Research* 53(5), 553. https://doi.org/10.1071/SR14297.

Tran, D. T. Q., Bradbury, M. I., Ogtrop, F. F. V., Bozkurt, H., Jones, B. J. and McConchie, R. 2020. Environmental drivers for persistence of Escherichia coli and Salmonella in manure-amended soils: a meta-analysis. *Journal of Food Protection* 83(7), 1268-1277. https://doi.org/10.4315/0362-028X.JFP-19-460.

Tsiafouli, M. A., Thébault, E., Sgardelis, S. P., de Ruiter, P. C., van der Putten, W. H., Birkhofer, K., Hemerik, L., de Vries, F. T., Bardgett, R. D., Brady, M. V., Bjornlund, L., Jørgensen, H. B., Christensen, S., Hertefeldt, T. D., Hotes, S., Gera Hol, W. H., Frouz, J., Liiri, M., Mortimer, S. R., Setälä, H., Tzanopoulos, J., Uteseny, K., Pižl, V., Stary, J., Wolters, V. and Hedlund, K. 2015. Intensive agriculture reduces soil biodiversity across Europe. *Global Change Biology* 21(2), 973-985. https://doi.org/10.1111/gcb.12752.

Tubeileh, A. M. and Stephenson, G. T. 2020. Soil amendment by composted plant wastes reduces the Verticillium dahliae abundance and changes soil chemical properties in a bell pepper cropping system. *Current Plant Biology* 22, 100148. https://doi.org/10.1016/j.cpb.2020.100148.

Vance, E. D., Brookes, P. C. and Jenkinson, D. S. 1987. An extraction method for measuring soil microbial biomass C. Soil Biology. *Soil Biology and Biochemistry* 19(6), 703-707. https://doi.org/10.1016/0038-0717(87)90052-6.

Verhoeven, E., Decock, C., Garland, G. and Lazcano, C. 2019. Vineyard nitrous oxide (N2O) emissions. Wine Business Monthly 10. Available at: (https://www.winebusiness.com/wbm/?go=getArticleSignIn&dataId=207657).

Wade, J., Hamilton, M. A., Beetstra, M. L., Culman, S. W. and Margenot, A. J. 2022. Soil health conceptualization differs across key stakeholder groups across the Midwest. *Journal of Soil and Water Conservation* 76(6), 1–7.

Wagg, C., Bender, S. F., Widmer, F. and van der Heijden, M. G. A. 2014. Soil biodiversity and soil community composition determine ecosystem multifunctionality. *Proceedings of the National Academy of Sciences of the United States of America* 111(14), 5266–5270. https://doi.org/10.1073/pnas.1320054111.

Walling, E. and Vaneeckhaute, C. 2020. Greenhouse gas emissions from inorganic and organic fertilizer production and use: a review of emission factors and their variability. *Journal of Environmental Management* 276, 111211. https://doi.org/10.1016/j.jenvman.2020.111211.

Wang, S. Y. and Lin, H. S. 2003. Compost as a soil supplement increases the level of antioxidant compounds and oxygen radical absorbance capacity in strawberries. *Journal of Agricultural and Food Chemistry* 51(23), 6844–6850. https://doi.org/10.1021/jf030196x.

Ward, D. M., Weller, R. and Bateson, M. M. 1990. 16S rRNA sequences reveal numerous uncultured microorganisms in a natural community. *Nature* 345(6270), 63–65. https://doi.org/10.1038/345063a0.

Watts, D. B., Torbert, H. A., Feng, Y. and Prior, S. A. 2010. Soil microbial community dynamics as influenced by composted dairy manure, soil properties, and landscape position. *Soil Science* 175(10), 474–486. https://doi.org/10.1097/SS.0b013e3181f7964f.

Weber, J., Karczewska, A., Drozd, J., Licznar, M., Licznar, S., Jamroz, E. and Kocowicz, A. 2007. Agricultural and ecological aspects of a sandy soil as affected by the application of municipal solid waste composts. *Soil Biology and Biochemistry* 39(6), 1294–1302. https://doi.org/10.1016/j.soilbio.2006.12.005.

Weil, R. R., Islam, K. R., Stine, M. A., Gruver, J. B. and Samson-Liebig, S. E. 2003. Estimating active carbon for soil quality assessment. A simplified method for laboratory and field use. *American Journal of Alternative Agriculture* 18(1), 3–17.

Weindorf, D. C., Zartman, R. E. and Allen, B. L. 2006. Effect of compost on soil properties in Dallas, Texas. *Compost Science and Utilization* 14(1), 59–67. https://doi.org/10.1080/1065657X.2006.10702264.

Weller, D. M., Raaijmakers, J. M., Gardener, B. B. M. and Thomashow, L. S. 2002. Microbial populations responsible for specific soil suppressiveness to plant pathogens. *Annual Review of Phytopathology* 40, 309–348. https://doi.org/10.1146/annurev.phyto.40.030402.110010.

Willekens, K., Vandecasteele, B., Buchan, D. and De Neve, S. 2014. Soil quality is positively affected by reduced tillage and compost in an intensive vegetable cropping system. *Applied Soil Ecology* 82, 61–71. https://doi.org/10.1016/j.apsoil.2014.05.009.

Willett, W., Rockström, J., Loken, B., Springmann, M., Lang, T., Vermeulen, S., Garnett, T., Tilman, D., DeClerck, F., Wood, A., Jonell, M., Clark, M., Gordon, L. J., Fanzo, J., Hawkes, C., Zurayk, R., Rivera, J. A., De Vries, W., Majele Sibanda, L., Afshin, A., Chaudhary, A., Herrero, M., Agustina, R., Branca, F., Lartey, A., Fan, S., Crona, B., Fox, E., Bignet, V., Troell, M., Lindahl, T., Singh, S., Cornell, S. E., Srinath Reddy, K., Narain, S., Nishtar, S. and Murray, C. J. L. 2019. Food in the Anthropocene: the EAT-Lancet commission on healthy diets from sustainable food systems. *The Lancet* 393(10170), 447–492. https://doi.org/10.1016/S0140-6736(18)31788-4.

Wind, L., Krometis, L. A., Hession, W. C. and Pruden, A. 2021. Cross-comparison of methods for quantifying antibiotic resistance in agricultural soils amended with dairy manure and compost. *Science of the Total Environment* 766, 144321. https://doi.org/10.1016/j.scitotenv.2020.144321.

Wolff, M. W., Alsina, M. M., Stockert, C. M., Khalsa, S. D. S. and Smart, D. R. 2018. Minimum tillage of a cover crop lowers net GWP and sequesters soil carbon in a California vineyard. *Soil and Tillage Research* 175, 244–254. https://doi.org/10.1016/j.still.2017.06.003.

Wortman, S. E. 2015. Crop physiological response to nutrient solution electrical conductivity and pH in an ebb-and-flow hydroponic system. *Scientia Horticulturae* 194, 34–42. https://doi.org/10.1016/j.scienta.2015.07.045.

Wortman, S. E., Holmes, A. A., Miernicki, E., Knoche, K. and Pittelkow, C. M. 2017. First-season crop yield response to organic soil amendments: a meta-analysis. *Agronomy Journal* 109(4), 1210–1217. https://doi.org/10.2134/agronj2016.10.0627.

Wu, W. and Ma, B. 2015. Integrated nutrient management (INM) for sustaining crop productivity and reducing environmental impact: a review. *Science of the Total Environment* 512–513, 415–427. https://doi.org/10.1016/j.scitotenv.2014.12.101.

Xin, X., Zhang, J., Zhu, A. and Zhang, C. 2016. Effects of long-term (23 years) mineral fertilizer and compost application on physical properties of fluvo-aquic soil in the North China Plain. *Soil and Tillage Research* 156, 166–172. https://doi.org/10.1016/j.still.2015.10.012.

Young, J. C. 1995. Microwave-assisted extraction of the fungal metabolite ergosterol and total fatty acids. *Journal of Agricultural and Food Chemistry* 43(11), 2904–2910. https://doi.org/10.1021/jf00059a025.

Zhang, Q., Miao, F., Wang, Z., Shen, Y. and Wang, G. 2017. Effects of long-term fertilization management practices on soil microbial biomass in China's cropland: a meta-analysis. *Agronomy Journal* 109(4), 1183–1195. https://doi.org/10.2134/agronj2016.09.0553.

Zhen, Z., Liu, H., Wang, N., Guo, L., Meng, J., Ding, N., Wu, G. and Jiang, G. 2014. Effects of manure compost application on soil microbial community diversity and soil microenvironments in a temperate cropland in China. *PLoS ONE* 9(10), e108555. https://doi.org/10.1371/journal.pone.0108555.

Zhu, X., Jackson, R. D., DeLucia, E. H., Tiedje, J. M. and Liang, C. 2020. The soil microbial carbon pump: from conceptual insights to empirical assessments. *Global Change Biology* 26(11), 6032–6039. https://doi.org/10.1111/gcb.15319.

Chapter 3

Optimizing slurry management

David Fangueiro, LEAF-Instituto Superior de Agronomia-ULisboa, Portugal; Jihane Elmahdi*, Wageningen University and Research, The Netherlands; Jared Nyang'au, Aarhus University, Denmark; Stamatis Chrysanthopoulos, LEAF-Instituto Superior de Agronomia-ULisboa, Portugal; Jerke De Vries, Wageningen University and Research, The Netherlands; and Peter Sørensen, Aarhus University, Denmark

1 Introduction
2 Current decision tools for optimizing manure management
3 Modifying animal slurry pH to enhance its value as a biobased fertilizer: (bio)-acidification and alkalinization
4 Improving manure management systems to minimize trade-offs
5 Combining manure management with anaerobic digestion
6 Pre- and post-treatment of biomass for anaerobic digestion
7 Optimization of anaerobic digestion operations to optimize digestate quality
8 References

1 Introduction

The rising demand for food has prompted an intensification of agricultural practices (Abdalla, 2002), leading to a revolution in food production with a focus on improving productivity (Wright et al., 2012; Eurostat, 2016). This has encouraged a growing reliance on mineral fertilizers (especially nitrogen (N), phosphorus (P) and potassium (K)). However, the extensive use of mineral fertilizers poses long-term environmental problems (Welch, 2002), including the greenhouse gases (GHG) involved in manufacturing synthetic fertilizers and the non-renewable nature of raw materials such as phosphate used to produce phosphorus (P) fertlizer (Eurostat, 2016). The European energy crisis has also had a significant impact on the availability and cost of synthetic fertilizers. In addition, these fertilizers do not contribute to levels of soil organic matter (SOM) which are increasingly seen as critical to long-term soil health and sustainable crop yields.

* Agricultural Biosystems Engineering Department, Wageningen University and Research, The Netherlands.

http://dx.doi.org/10.19103/AS.2023.0120.12

Animal manure is seen a potential solution to address the dual challenges of replacing synthetic fertilizers in optimizing crop yields whilst also building up soil organic matter at nominal expense. Animal manure includes not only nutrients but also organic matter (OM) and beneficial microorganisms that can improve soil health and thus crop yields (Köninger et al., 2021). This approach also aligns with recent EU strategies promoting recycling and a circular economy.

However, animal manures are also responsible for close to 10% of global GHG emissions and are the main source of ammonia emissions. If not properly managed, they can also negatively impact water quality due to N and P leaching and run-off from fields into watercourses. Global annual production of livestock manure is estimated to be about 80-140 Tg, with a N content exceeding synthetic N production (Oenema and Tamminga, 2005; Zhang et al., 2020). Cattle manure has the largest share at about 60% of total manure (Thangarajan et al., 2013). It is estimated that 50% of manure is deposited in pastures while 50% is collected: only half of the amount collected is recycled to agricultural land. Manure is one of the main sources of nitrous oxide (N_2O) and methane (CH_4) emissions in agriculture (Burney et al., 2010; Zhang et al., 2020). It has been estimated that, of the total N excreted by animals, only 15% is taken up by any subsequent crop (with a range of 0-60%). The other 85% of excreted N is lost via NH_3 volatilization, nitrification-denitrification, leaching and run-off (Oenema and Tamminga, 2005; Zhang et al., 2020). Manure also contains phosphorus that can be lost after soil application, resulting in aquatic eutrophication (Kleinman et al., 2020). These losses of P and N not only waste agronomically valuable nutrients but also contribute to serious environmental impacts such as climate change, soil acidification, eutrophication and harmful particulate matter formation in the atmosphere (Yuan et al., 2018).

Despite many advances in research, there is still a need for new, improved strategies for manure management to optimize its use and minimize emissions. One issue is a lack of a clear definition of manure in EU regulations: from 'waste products excreted by livestock' in the Nitrate directive (Directive 91/676/EEC) to 'any excrement and/or urine of farmed animals (...), with or without litter' in the Regulation on animal by-products (EC/1069/2009). Another factor to consider is that up to 90% of manure applied to soils has no or minimal processing (Foged et al., 2011).

Part of the reason for the lack of manure processing is the strong imbalance in production of manure in regions such as Europe (Eurostat, 2021). Particular areas in countries like Germany, Spain, France or Netherlands are hotspots of manure production. There has been a growing specialization of crop and livestock production with livestock production generally concentrated in specific areas and on specific livestock farms (sometimes operating on a large scale). In 2018, e.g. 80% of manure was produced by only 4% of European farms (Amann et al., 2018). This means both a problem for those livestock farms having to manage very large volumes of manure and for the wider agricultural sector in managing how this resource can be efficiently distributed to benefit crop production. This means that dealing with manure is as much about management as it is about technical solutions for treating manure (Niles et al., 2019; Varma et al., 2021). This is particularly the case given the risk of pollution swapping where, in solving one pollution problem, a particular treatment unintentionaly creates a new environmental problem. Indeed, techniques such as anaerobic digestion (AD), the use of covers for storage, slurry injection and acidification can all potentially result in pollution swapping (Emmerling et al., 2020). This highlights the importance of an integrated management approach.

These challenges necessitate the adoption of appropriate techniques and strategies for manure management to optimize the recovery of nutrients, mitigate the environmental consequences associated with its storage, transportation and application on soil and ensure its safe use (Amon et al., 2021; Fangueiro et al., 2023). This chapter examines how to achieve these objectives by looking at options at the various stages involved in slurry management, including collecting excreta within the animal house, storage, processing and field application. Each of these steps involves a multitude of potential treatments and technologies that can be employed to enhance use in terms of the agronomic/nutrient value of manure products reducing GHG emissions, odours and cost.

2 Current decision tools for optimizing manure management

A large number of tools are available for manure management. A list of the most relevant decision tools is shown in Table 1. These tools have been divided in four categories:

- Manure production and characteristics;
- Manure storage;
- Soil application of manure; and
- Managing the whole manure management chain.

In the first category, there are three examples of decision tools to estimate the amount of manure produced and/or the amount of nutrients contained in manure:

- Nutrient loading calculator calculates the amount of N, P, K and S that will be deposited in the form of manure, urine and waste feed depending on

Table 1 Most relevant tools available for manure management at several steps of the manure management continuum

Category	Name of tool	Link	Authors
Manure production and characteristics	Nutrient Loading Calculator	https://www.alberta.ca/nutrient-loading-calculator.aspx	Alberta government and Agri-Food Canada in collaboration with prairie provincial agricultural departments
	Manure Properties Calculation Tool	https://projects.luke.fi/manurestandards/en/results/	Project Manure Standards: Allan Kaasik, Christian Friis Børsting, Friederike Lehn
	COMPOSIM	http://www.rmt-fertilisationetenvironnement.org/moodle/mod/resource/view.php?id=802	Institut de l'élevage
Manure storage	Le Pré-Dexel et le Dexel	https://idele.fr/gestion-des-effluents-et-des-dejections-ged?tx_ideleregeff_feregeff%5Baction%5D=displayRessources&tx_ideleregeff_feregeff%5BcomeFrom%5D=themes&tx_ideleregeff_feregeff%5Bcontroller%5D=Screen&tx_ideleregeff_feregeff%5BwannaGoTo%5D=REGLEMENTATION_ZONE_VULNERABLE&cHash=4cb71ef3249f10af4d75226e3a5b9465	Institut de l'élevage
Soil application of manure	Manure Transportation Calculator	https://www.alberta.ca/manure-transportation-calculator.aspx	Alberta government
	Phosphorus Management Tool	https://www.alberta.ca/phosphorus-management-tool.aspx	developed by the Government of Alberta as part of the Alberta Phosphorus Watershed Project
	Manure Management Planner	https://www.alberta.ca/manure-management-planner.aspx	Purdue University for use in the United States. The programme was adapted by the Alberta government for use in Alberta

Category	Tool	Reference / URL	Provider
	Sorensen model	Sørensen, C.G., Jacobsen, B.H., Sommer, S.G., 2003. An Assessment Tool applied to Manure Management Systems using Innovative Technologies, Biosystems Engineering, 86, 315-325 https://doi.org/10.1016/S1537-5110(03)00137-5.	Claus G. Sørensen, Brian H. Jacobsen, Sven G. Sommer
	BeMestWijs management tool: Smart manure application and utilization:	https://www.wur.nl/en/article/Smart-manure-application-utilisation-a-practical-tool-for-issuing-advice-on-manure-application.htm	ir. G (Gerard) Migchels (wageningen) LTO Bedrijven LTO Noord DairyCampus
	Deere calculator coupled with HarvestLab 3000	https://www.deere.co.uk/en/tools/ag-turf/manure-sensing-calculator/	John Deer Company
	Ermes	https://www.ermes.pro/fonctionnalitesgenerales.html	IG tools
	EpandApp	https://www.mon-cultivar-elevage.com/content/regler-son-epandeur-sur-smartphone-avec-epandapp	Chambre Régionale d'Agriculture de Bretagne
	Manure Tracker	https://play.google.com/store/apps/details?id=ipcm.counter-prelim2&hl=pt_PT&gl=US	University of Wisconsin Integrated Pest Management
	Manure Valuator app	https://www.manuremanager.com/manure-app-calculates-value-14819/	University of Arkansas Division of Agriculture
Whole manure management chain	The Farm Crap App Pro	https://www.agricology.co.uk/resources/farm-crap-app-pro	SWARM Hub
	Manure Management N-flow tool	https://www.eea.europa.eu/publications/emep-eea-guidebook-2019/part-b-sectoral-guidance-chapters/4-agriculture/manure-management-n-flow-tool/view	Aether Ltd 2019 under contract to the EEA
	Manure Monitor	https://play.google.com/store/apps/details?id=com.movecreative.feedlotandroid&hl=pt_PT&gl=US	Move Creative

the feeding strategy considered. It makes it possible to adapt the herd to the available feed or the amount of feed needed for the existing herd.

- Manure Properties Calculation Tool allows the estimation of manure composition.
- COMPOSIM simulates the amount and characteristics of manure produced.

In the area of manure storage, there are two French decision tools:

- Pré-Dexel; and
- Dexel.

These make it possible to estimate the volume of storage needed in a specific farm and specifically the storage capacity needed to conform to the rules in nitrate vulnerable zones.

There are a large number of tools dedicated to the application of manure to soil. These include

- manure management planner; and
- BeMestWijs management tool.

These define where, when and how much manure can be applied. Both make it possible to simulate the effects of improvements/changes in management. Other tools include:

- Manure management planner: This analyses if it is possible to apply all the manure produced on a farm and enables long-term manure planning for up to 10 years.
- BeMestWijs management tool: This uses data from a number of management tools and databases to allow precision manure application.
- Manure transportation calculator: This analyses the costs associated with transport and application of manure.
- Deere calculator coupled with HarvestLab 3000: This allows variable-rate application of slurry based on continuous analysis of the composition of slurry using NIRS.
- Phosphorus management tool: This assesses the impact of manure application on P losses; it also gives information on best management practices to decrease P losses.
- ERMES: This uses geolocalization to track the location and amount of manure applied; an excellent tool to prepare manure application maps and assess the impacts of cumulative applications.

More recently, tools have been created as Apps to allow farmers and others easy access via smartphones. Among several Apps, we have selected four different tools with distinct objectives:

- Farm crap App determines the amount of crop-available nutrients (nitrogen (N), phosphorous (P) and potassium (K)) for different spreading rates of manure. This App helps to establish how much to spread in order to meet crop requirements with minimum waste, environmental impact and cost.
- Manure Valuator App calculates the value of manure based on the amount of fertilizer it replaces.
- Manure Tracker App automatically records manure application on fields using GPS (complementing the ERMES tool).
- Epand calculates the amount of manure to apply to reach an N target and can calibrate machinery used for manure application.

As noted earlier, avoiding problems such as pollution swapping requires a holistic approach. A final set of tools considers the whole manure management chain. These include:

- Manure management N-flow tool looks at N flows with special emphasis on GHG emissions (Sorensen et al., 2003). It compares different manure management strategies to identify the best solutions for specific farms. It also enables a system-oriented evaluation of labour requirements, machinery capacity and costs related to manure handling for any one approach.
- Manure Monitor is an App created to facilitate the recording of data related to manure management; it also includes an emergency response plan in case of manure-related pollution incidents.

3 Modifying animal slurry pH to enhance its value as a biobased fertilizer: (bio)-acidification and alkalinization

With over 140 million animals, the European Union is the second largest global producer of pigs, generating large quantities of slurry (Eurostat, 2022). Limited storage capacity in pig-breeding facilities and lack of a uniform manure regulation have resulted in untreated pig slurry being spread on fields. This practise has been associated with serious potential safety and environmental impacts, particularly in ammonia (NH_3) emissions (Hjorth et al., 2010), nutrient leaching to water courses (Fangueiro et al., 2014) and potential contamination of crops with pathogens (Rodrigues et al., 2021).

A promising and cost-effective approach is pH modification of swine slurry using additives. Due to its low dry matter content, slurry can be seen as an aqueous solution where an equilibrium exists between ammonium (NH_4^+) and unionized NH_3. This equilibrium is pH dependent and favours NH_4^+

concentration once animal slurry is acidified (Regueiro et al., 2016a). Modifying the pH of animal slurry can therefore cause significant changes to its physico-chemical and biological properties (Fangueiro et al., 2015). A typical range of initial animal manure pH lies between 7 and 8.4 (Regueiro et al., 2016). At this range, buffer components (weak acids or bases) are responsible for pH regulation. Slurry can exhibit high buffer capacity which can affect the amount of additive required to modify the pH (Sommer and Husted, 1995) .

A key benefit in altering slurry pH is mitigation of nitrogen (N) losses to the atmosphere via gaseous NH_3 (Regueiro et al., 2016; Overmeyer et al., 2021). Since NH_3 is a nitrogenous compound, the N fertilizer value of the slurry is also increased. Slurry sanitization, a crucial aspect related to the safety of biobased fertilizers, can also be attained through pH modification (Rodrigues et al., 2021). Alkalinization followed by stripping technologies also allows recovery of ammonia and the production of ammonium-based fertilizers (Overmeyer et al., 2020). There are, however, problems to overcome, including replacing inorganic with organic acid/alkali compounds to allow treated slurry to be used in organic agriculture (see European Regulation 2019/1009) and effective integration of this technique with others in an overall manure management strategy (Hjorth et al., 2010; Varma et al., 2021).

There are three strategies to modify the pH of animal slurry:

- Chemical acidification (Fangueiro et al., 2015);
- Biological acidification (Regueiro et al., 2022); and
- Alkalinization (Rodrigues et al., 2021).

Acidification is the most common strategy in reducing slurry pH. Acidification of slurry with inorganic acids, such as sulphuric acid (H_2SO_4), has been practised at a farm scale in Northern European countries (i.e. Denmark) for almost 20 years (Fangueiro et al. 2015; Regueiro et al., 2022). As alternatives to inorganic acids, organic acids (like acetic acid and citric acid) have also been assessed for their efficacy to reduce pH (Regueiro et al., 2016). A typical target pH is 5.5 since, at this range, the NH_4^+/NH_3 equilibrium favours NH_4^+ (Overmeyer et al., 2021).

Fangueiro et al. (2015) reported up to 70% and 88% NH_3 reduction with H_2SO_4 during in-house and storage acidification, respectively. Methane (CH_4) emissions from stored slurry can also be reduced by up to 99% by reducing its pH value to 5.5 (Fuchs et al., 2021). The fertilizer value of acidified slurry exhibits higher concentration of dissolved nutrients (Fangueiro et al., 2015). For instance, the solubility of phosphorus (P), e.g. is increased by acidification of cattle slurry (Pedersen et al., 2017). Swine slurry acidification to pH 5.5, e.g. increased the dissolution of total P (~70% as Ortho-P) (Christensen et al., 2009), facilitating P uptake by crops such as maize (Regueiro et al., 2020). Acidification is also important in slurry sanitization (Köninger et al., 2021). The EU regulation for fertilizers ((EU) 2019/1009) mandates a limit of 1000 CFU/g in five out of five tested samples for *Escherichia coli* or total coliforms, and the absence of *Salmonella* spp. in 25 g of fertilizer. Rodrigues et al. (2021) found effective swine slurry sanitization (<1000 CFU *E. coli*) was possible with H_2SO_4 at a target pH of 5.

There are, however, disadvantages with chemical acidification. The higher nutrient solubility that acidified slurry exhibits can increase nutrient leaching (Fangueiro et al., 2014). While acidification abates NH_3 losses, there are trade-offs in terms of N_2O emissions (Schreiber et al., 2023). In a study where untreated and acidified slurry were applied at a rate of 90 kg N ha^{-1}, 35% higher cumulative soil N_2O emissions were observed with acidified slurry compared to untreated slurry (Malique et al., 2021). A meta-analysis by Emmerling et al. (2020), however, concluded a decrease in CH_4, CO_2 and N_2O emissions from acidified slurry. Concentrated H_2SO_4 also poses significant safety risks for workers and is challenging to handle (Bastami et al., 2016; Prado et al., 2020). Cost is also an issue. de Vrieze et al. (2019) concluded that acidification with H_2SO_4 increased total net process costs by approximately 24%.

Biological-acidification (bio-acidification) of animal slurry is a microbially-driven process where the microorganisms present in slurry convert easily fermentable carbon sources into organic acids such as lactic acid (Regueiro et al., 2022). Bio-acidification with glucose and sucrose as substrates has exhibited promising results in pH modification and reduction in GHGs from animal slurry (Bastami et al., 2016; Prado et al., 2020; Regueiro et al., 2022). Bastami et al. (2016), e.g. observed CH_4 reduction of up to 99% using brewing sugar as labile C source to bio-acidify dairy slurry. Bio-acidification is able to achieve a target pH of 5.5 or lower (Prado et al., 2020). However, high doses of sucrose-rich additives are required to initiate and maintain the fermentation process, and this may not be economically viable since sucrose is often more valuable for other applications. Combining slurry acidification using H_2SO_4 followed by the addition of glucose and/or sucrose-rich additives has been explored as an alternative (Overmeyer et al., 2021; Regueiro et al., 2022). Pre-acidification of swine slurry with H_2SO_4 to pH 5.5 followed by glucose addition (2% and 4%) appeared to maintain pH below 5 more consistently than glucose (2%) treatment alone (Regueiro et al., 2022).

Slurry alkalinization, a strategy much less studied, is the opposite of acidification since it increases slurry pH. Additives used to increase animal slurry pH include lime products such as calcium hydroxide ($[Ca(OH)_2]$) as well as other hydroxide salts like potassium hydroxide (KOH) and sodium hydroxide (NaOH) (Rodrigues et al., 2021). In principle, alkalinization shifts the NH_4^+/NH_3 equilibrium towards the latter, causing NH_3 losses to the atmosphere and consequently reducing the fertilizer value of the slurry. However, alkalinization at pH 12 can inactivate pathogenic microorganisms (Anderson et al., 2015).

Table 2 Agro-industrial by-products as alternative additives to modify the pH of animal slurry

Additive	Slurry type used	Tested amount	Additional benefit	References
Apple pulp Brewers grain Dairy washings Dairy waste Grass silage Maize silage Sugar beet molasses	Dairy cattle slurry	70;150 70;150 150 100 70;150 70;150 30;50;70 (g kg^{-1})	Reduce NH_3 (up to 67%) and CH_4 (15-70%) emissions during storage	(Kavanagh et al., 2021)
Brewing sugar Brewing sugar & Actiferm EM	Dairy cattle slurry	10% (w.w.) 10% & 5% (v.w.)	Reduce CH_4 (85-99%) emissions during storage	(Bastami et al., 2016)
Brown juice (coagulated & fermented; different combinations with H_2SO_4 and glucose)	Pig slurry	20-50 % (w.w.)	pH reduction up to 3.9 during storage Increased lactic acid production with fermented brown juice Negligible CH_4 and CO_2 emissions	(Regueiro et al., 2022)
Cheese whey Rice bran	Dairy slurry	100-250 (g 0.5 kg^{-1} slurry)	Reduce NH_3 emissions (58%) with cheese whey P and Mg enrichment in slurry treated with rice bran	(Prado et al., 2020)
Acidic whey Flushing milk	Cattle manure slurry	20-50% (v.w.)	pH reduction during storage (1 pH unit for flushing milk and up to pH 4.5 with acidic whey)	(Sepperer et al., 2021)
Banana peel	Pig slurry	20-120 (g L^{-1} slurry)	pH reduction up to 5.4 Dissolution of particulate P	(Moyo et al., 2022)
Cheese whey	Dairy cattle slurry	10-80% (w.w.)	pH reduction below 5 during storage Reduce total GHG's (up to 54%) and NH_3 (up to 90%) during storage Increase CH_4 (up to 53%) production during AD	(Gioelli et al., 2022)

Rodrigues et al. (2021) succeeded in sanitizing pig slurry with KOH addition by reaching pH 9.5, though they observed higher NH_3 emissions compared to acidified slurry. Alkalinization has also been tested in dairy sludge and anaerobically digested biosolids (Case and Jensen, 2019).

An alternative approach is to use agro-industrial by-products as additives to reach a target pH and enhance the value of slurry as biobased fertilizer (Chojnacka et al., 2020; Prado et al., 2020). Table 2 summarizes the range of alternative additives that have been used to modify slurry pH, primarily via bio-acidification with a particular focus on reducing GHGs and NH_3 emissions.

Acidified manure has also been considered as a potential feedstock in AD to produce energy (biogas) from organic residues such as animal manure (Chojnacka et al., 2020). Two main approaches have been considered:

- Acidification of raw slurry prior to AD; and
- Co-digestion of raw and acidified slurry.

Using the first approach, Sutaryo et al. (2013) acidified raw sow slurry with H_2SO_4 to pH 5.5. They observed a significant reduction in CH_4 yields compared to untreated slurry (256.6 L kg VS^{-1} and 372.7 L kg VS^{-1}, respectively). This was attributed to the inhibition of sulphur reduction in the acidified slurry. However, utilization in AD of cattle slurry bio-acidified with cheese whey resulted in a significant increase in CH_4 production (+53%) compared to non-acidified slurry (Gioelli et al., 2022). The feedstock ratio between acidified and non-acidified slurry fraction is crucial in the second approach. Moset et al. (2016) found that the 10% inclusion of acidified slurry in a thermophilic reactor increased CH_4 yields by 10%; however, CH_4 yields were reduced with over 30% acidified cattle slurry. Co-digestion of acidified pig slurry (30% inclusion) reduced CH_4 yields (Moset et al., 2012, 2016).

Solid-liquid separation of manure is a common treatment on livestock farms (Varma et al., 2021). However, solid and liquid fractions can release much higher NH_3 and GHG emissions compared to non-separated raw slurry during storage (Prado et al., 2022). This requires acidification of slurry prior to solid-liquid separation, which can also increase P concentration in the acidified liquid fraction (Fangueiro et al., 2009). Acidification of a separated liquid fraction to pH 5.5 required less H_2SO_4 compared to raw slurry but resulted in significant pH increase during storage (Overmeyer et al., 2021). de Vrieze et al. (2019) evaluated 15 different methods to enhance P recovery, of which nine involved acidification of swine manure or digestate with H_2SO_4. Most of these scenarios involved manure acidification prior to centrifugation to assist P transfer from the solid to the liquid fraction.

In conclusion, pH modification of pig and other types of slurry has both strengths and weaknesses. Slurry acidification can reduce NH_3 and CH_4

emissions, though pollution swapping is possible, particularly with N_2O emissions. Little attention has been paid on increasing slurry's initial pH, probably due to an expected increase in NH_3 emissions. The fate of pathogens and nutrient enrichment of slurry (when using KOH as additive) are strong arguments to explore alkalinization further. A recent development is combining additives to mitigate the use of H_2SO_4 (Fuchs et al., 2021). Alternative additives from agro-industrial activities could replace chemical acids/bases but more research is needed to evaluate their impact on the environment, crop fertilization and economic feasibility. Integration of slurry acidification in a manure management system requires careful design, including economic assessment for large-scale application.

4 Improving manure management systems to minimize trade-offs

A main reason behind failure of current technologies to effectively limit all environmental impact is the trade-offs taking place between different emissions. This requires a holistic manure management strategy that considers key aspects like emissions, nutrient availability and practicability for farmers.

Measures to reduce environmental impacts can be implemented at various steps of the manure management chain. Reducing crude protein in diets, e.g. can reduce NH_3 emissions in animal housing, during storage and following soil application (Hou et al., 2015). Solid-liquid separation has been found to effectively reduce CH_4 and N_2O emissions (Fangueiro et al., 2008; Aguirre-Villegas et al., 2014; Aguirre-Villegas et al., 2019; Emmerling et al., 2020; Kupper et al., 2020). However, separation and emission reduction technologies can require costly and complex technologies and may not deal with emissions before separation (Vu et al., 2016).

Source separation is gaining interest based on simple principles like gravity to separate urine from faeces in animal houses (Aarnink et al., 2007; De Vries et al., 2013; Koger et al., 2014; Loussouarn et al., 2014; Vu et al., 2016). In principle, separating faeces and urine at source should ensure rapid removal of faeces from the animal house, separate urea (in urine) from urease (faeces), and separate ammonium (in urine) from a carbon-rich source (faeces). This could potentially reduce emissions of NH_3, N_2O and CH_4 (Fig. 1) simultaneously. Compared to traditional slurry systems, source separation in pig housing systems was found to decrease NH_3 emissions by 46 to 70% and CH_4 emissions by up to 80% (Aarnink et al., 2007; Aarnink and Ogink, 2007; De Vries et al., 2015b). A source separation system can also achieve high separation efficiencies in terms of nutrients compared to traditional liquid/solid separation from slurry (mainly N from P) (Vu et al., 2016). Since most research is on pigs, more research is needed on source separation systems for dairy manure.

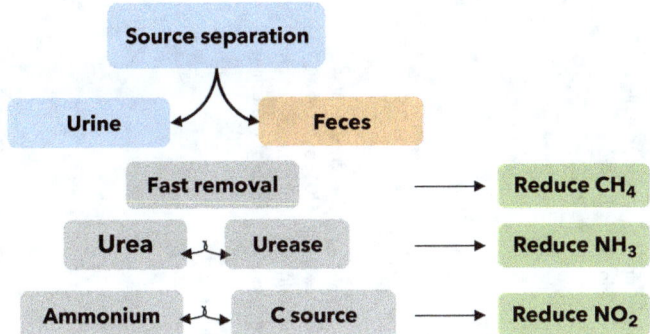

Figure 1 Potential effect of source separation of manure on emissions and involved mechanisms. C refers to carbon.

Emissions during storage are a particular concern. Strategies to manage emissions (and, ideally, valorize slurry) include AD, acidification, composting, drying, use of different natural additives (zeolite, biochar and straw) and storage covers. AD has been found to reduce CH_4 emissions during storage and alter nutrient availability (increase mineral N and reduce soluble P) (Clemens et al., 2006; Chadwick et al., 2011 ; De Vries et al., 2012; Hou et al., 2015; Dennehy et al., 2017 ; Hou et al., 2018 ; Maldaner et al., 2018 ; Aguirre-Villegas et al., 2019; Cao et al., 2019 ; Ershadi et al., 2020 ; Li et al., 2020). Acidification and use of covers are effective tools to reduce NH_3 emissions during storage and following field application (Ten Hoeve et al., 2016; Hou et al., 2017; Yuan et al., 2018; Fangueiro et al., 2015; Emmerling et al., 2020; Wagner et al., 2021; Keskinen et al., 2022). Better application methods and timing, and avoiding disturbance after application, can mitigate gaseous emissions and nutrient leaching (Webb et al., 2010; Chadwick et al., 2011; Emmerling et al., 2020; Erwiha et al., 2020; Dittmer et al., 2020; De Vries et al., 2015b). However, as noted earlier, there are usually trade-offs in using any one measure which need to be addressed by a more integrated approach.

Most studies of integrated systems using a range of techniques focused on limiting trade-offs between NH_3 and GHG emissions, or between environmental impacts like global warming, acidification and eutrophication (Aguirre-Villegas et al., 2014; Holly et al., 2017; Yuan et al., 2018; Aguirre-Villegas et al., 2019). These studies quantify NH_3 and GHG emissions at different stages of the chain, and sometimes energy used in manure processing and nutrient availability for crops.

Aguirre-Villegas et al. (2014), e.g. modelled the effect of solid-liquid separation of slurry, AD and both techniques combined, on NH_3, global warming potential, fossil fuel depletion and nitrogen availability from a whole manure management chain (barn, storage and field application). Slurry was stored in open basin and applied to field using broadcasting. AD combined

Table 3 Function of the single strategies, their pollution swapping and related problems, and proposed integrated solutions

Established single strategies	Function	Problem (pollution swapping)	Proposed integrated solution	References
Liquid/solid separation	-Manage nutrients -Easier handling -Reduce NH_3 and GHG emissions during storage	-Emissions in the animal house not tackled -Poor separation of nutrients (P and N stay in both fractions) -Soluble organic matter is available in liquid fraction causing CH_4 emission -Increase emissions of NH_3 and N_2O after field application especially from the liquid fraction	-Source separation (faeces/urine) separates nutrients efficiently and at the same time reduces NH_3, N_2O and CH_4 in the animal house	De Vries et al., 2013; Kupper et al., 2020; Emmerling et al., 2020
Anaerobic digestion (AD)	-Harvest CH_4/produce renewable energy reduces uncontrolled CH_4 emissions -Increase nutrient availability for plants (N and P)	-Increases emissions of NH_3 during storage and N_2O, NH_3, and NO_3^- after field application -Mono-digesting slurry is not cost effective	-Combining AD with drying of digestate or zeolite addition or acidification with sealed storage and adapted field application will control the emissions in all the stages after AD -Faster AD process of source separated faeces & possible higher yield per ton faeces -AD of source segregated faeces instead of slurry will need less digester size leading to reduced costs	De Vries et al., 2015a; Vu et al., 2016; Holly et al., 2017; Emmerling et al., 2020

Drying of slurry, manure digestates	-Reduce microbial activity leading to reduced emissions in storage -Easier handling/transport	-Massive NH_3 emission during drying -High energy required for slurry drying	-Combining with zeolite addition before drying and sealed storage will control the NH_3 emissions -Combining with source separation will control ammoniacal nitrogen loss because most ammoniacal nitrogen stays in the urine. -Combining with AD will provide renewable energy for drying -Drying faeces only will require less energy than slurry (lower volume and lower moisture content)	Liu et al., 2019; Aguirre-Villegas et al., 2019; De Vries et al., 2015a
Covered storage	-Reduce emissions during storage	-High emissions prior to storage (animal house) -N_2O and CH_4 emissions upon exposure in addition to NH_3 and after field application	-Combining with source separation will control the emissions in the animal house -Combining with AD will reduce CH_4 production during storage and volatilization after field application -Acidification of separated faeces and urine to pH 5.5 will control NH_3 emissions after field application -Combining with adapted application methods will control N_2O after field application	De Vries et al., 2015a; Yuan et al., 2018; Keskinen et al., 2022; Wagner et al., 2021
Injection application field	-Reduce NH_3 emissions	-Increase N_2O emissions -Excess N or P applied leading to leaching and run-off (unbalanced N/P ratio on animal manures)	-Combining with source separation, AD, additives or drying, and sealed outside storage will lead to separated treated urine and faeces with more balanced N/P ratios and less prone to emissions -Combining with adapted application rate, time and reduced intervention will result in the control of all emissions and leaching	Nyord et al., 2013; De Vries et al., 2015a; Wagner et al., 2021; Keskinen et al., 2022

with solid-liquid separation had a lower impact in studied categories except NH_3 emissions. NH_3 emissions increased by 44% compared to untreated slurry, whilst a 40% and 2% increase occurred with single use of AD and solid-liquid separation, respectively. The combination of AD with solid-liquid separation also reduced nitrogen availability more than AD and solid-liquid separation used separately. These technologies are known to reduce GHG emissions but increase NH_3 emissions (Holly et al., 2017; Emmerling et al., 2020). An optimized management system thus needs to incorporate a NH_3 emissions mitigation strategy. Aguirre-Villegas et al. (2019) found AD combined with liquid-solid separation achieved a higher reduction of GHGs during storage than each of the technologies used separately. However, N_2O emissions during field application increased net GHG emissions.

Yuan et al. (2018) compared four pig slurry management systems integrating AD, separation and wastewater treatment and field application using life cycle assessment (LCA). Outdoor storage was uncovered and the slurry was surface applied. Results showed that even the best-performing system, which consisted of in-house separation and field application, still had high environmental impacts, in eutrophication and acidification. This study also highlights the importance of designing management systems considering all objectives and including interventions from the animal house to the field. Including techniques for optimized storage and field application could have reduced eutrophication and acidification potential through reduced NH_3 emissions. This study also suggested that the best management system could be further optimized by improving P removal, reducing P losses after soil application and eutrophication potential.

Table 3 summarizes potential ways to integrate treatment technologies to reduce all emissions simultaneously with minimum energy required. For each technology, the pollution-swapping problem is highlighted, and an integrated approach is proposed.

As noted, manure management involves potentially challenging trade-offs between environment, economic, community and farmer priorities (Niles et al., 2022). Integrated manure management systems would be more effective in balancing these trade-offs if they included an initial system design step. De Vries et al. (2015a) used engineering design methods to develop an integrated system covering the whole chain, including setting clear objectives and system boundaries to prevent pollution swapping and reduce the environmental impact of the whole chain. They highlighted the importance of starting with source segregation of urine and faeces since this was found to have no pollution swapping effect, could significantly reduce GHG emissions (up to 82%), acidification and particulate matter formation (up to 49%) with no significant effect on eutrophication (De Vries et al., 2013). When they modelled the effect of such an integrated system compared to conventional management, they

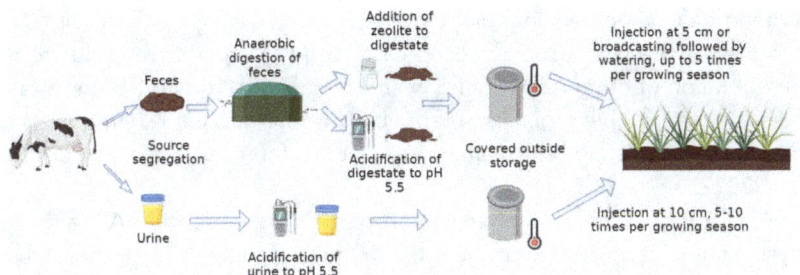

Figure 2 Graphical illustration of an integrated system to manage dairy manure from animal house to field.

found that an increase of up to 70% in N use efficiency, a decrease of excess P application to soil, and a reduction of GHG and NH_3 emissions with minimal energy use (De Vries et al., 2015b).

Steps to limit CH_4 emissions include the following:

1 Define factors regulating methane emissions in manure management.
2 Define functions that can be designed into the system to control these factors (C availability, temperature, oxygen and pH).
3 Define how these factors can be managed (e.g. remove C, reduce temperature, increase oxygen and lower pH).
4 Define techniques to achieve this (e.g. grooved floor, mono-digestion and adding acid).
5 Define possible interactions of this technique with other process, e.g. dealing with NH_3 or N_2O emissions.
6 Iterate between these steps.

The final result should be an integrated manure management system covering the whole chain (from the animal house up to field application). Figure 2 illustrates an example of an integrated system. A good integrated system is also one that can be verified by appropriate measurement and testing to follow manure fractions through all the steps in the chain. Measurement should be able to track changes in composition after each step/treatment and measure all emissions in the animal house, during storage and following field application.

5   Combining manure management with anaerobic digestion

One strategy to optimize manure management is to combine it with AD. AD is a bioenergy technique in which organic materials are broken down in an oxygen-free environment into biogas and digestate as a by-product. Digestate from the AD process is characterized by a high NH_4^+/total N ratio due to the mineralization of organically bound N and is stabilized due to the conversion of less decomposable organic matter to biogas. This reduces N immobilization after manure application and increases available N for plant utilization (Sørensen et al., 2012).

Substrate physical and chemical properties change during AD, affecting physico-chemical properties such as the dry matter content, NH_4^+/total N ratio, viscosity and pH (Barampouti et al., 2020). Digestate dry matter and viscosity decreases, enhancing soil infiltration. Low dry matter reduces contact time with the soil surface, reducing ammonia volatilization and improving nutrient value of the digestate (Hafner et al., 2018). Decomposition and recovery of labile carbon from substrates as biogas (CH_4 and CO_2) reduces N immobilization and increases plant-available nitrogen (Webb et al., 2013).

Organic matter content in manure is degraded during complex biochemical reactions in AD, resulting in thinner manure with better soil infiltration and less microbial immobilization of nutrients in the soil. Volatile solids in manure are recovered as CH_4 and CO_2 during AD, avoiding GHG emissions. Stabilized digestates decrease the microbial demand for oxygen after application, reducing heterotrophic denitrification and N_2O emissions (Chadwick et al., 2011). However, the AD increases manure pH, increasing the risk of ammonia volatilization (Sommer and Husted, 1995). This requires measures during storage and digestate application to minimize this risk.

Input substrate characteristics influence biogas yield and digestate quality (Hagos et al., 2017). The AD process may involve mono-digestion with only one substrate or co-digestion with multiple substrates digested together. Due to prior carbon degradation in the animal's digestive system, mono-digestion of manure for biogas production yields relatively low amounts of biogas. Mono-digestion is also characterized by nutritional imbalance and a lack of diverse microorganisms required to optimize the AD process (Hagos et al., 2017).

Co-digestion of manure and other substrates, such as food waste, industrial food wastes and agricultural crop residues, maximizes biogas yield (Moset et al., 2017; Sarker et al., 2019). It allows system optimization and improved economic viability of biogas plants, as well as the ability to recycle and reuse nutrients from other waste streams (Sayara and Sánchez, 2019; Sørensen et al., 2019). As a result, there has been an increasing shift towards co-digestion of the slurry with solid organic biomass (such as straw-rich litter, grass and straw) to produce biogas, for instance in Denmark (Møller et al., 2022).

The co-digestion of manure with solid biomasses has a high biogas potential. However, most substrates are lignocellulosic with high fibre content, limiting their degradation during AD. Digestion of recalcitrant biomass poses problems with pumping and agitation, which require investments in adapting biogas plants to handle the high dry matter content in solid biomass (Møller and Nielsen, 2016). These limitations result in low biogas yields and affect digestate characteristics such as dry matter content which influence nitrogen use efficiency during field application. These challenges require adapting AD systems.

Co-digestion of lignocellulosic biomasses with livestock slurry/manure results in a digestate with high dry matter content and viscosity due to the low digestibility of the lignin (Barampouti et al., 2020). This poses agronomic and environmental challenges during digestate storage and application (Olesen et al., 2021; Webb et al., 2013). High dry matter content increases the ammonia volatilization per unit area during surface application given the reduced infiltration rate of digestate (Søgaard et al., 2002). Losses are further exacerbated by high digestate pH which reduces fertilizer value. Møller et al. (2022) estimated an increase in ammonia emissions of 0.14 to 0.3 kg NH_3-N ton^{-1} biomass digested due to higher pH. Nyang'au et al. (2023) also reported higher ammonia volatilization from digestates with higher dry matter content following application to winter wheat compared to digestates with low dry matter content, which they attributed to reduced infiltration rates and higher pH.

Olesen et al. (2021) and Pedersen et al. (2021) found that adhesive characteristics and increased surface area of digestate might limit soil infiltration compared to untreated manure, resulting in higher NH_3 volatilization, outweighing the benefits of reduced dry matter content. NH_3 losses from digestates could be minimized by utilizing low-emission application techniques such as digestate injection into soil (Møller et al., 2022; Sørensen et al., 2019). However, this is difficult in some crops, such as in winter cereals, where injection can cause significant damage to crops. Alternative technologies, such as solid-liquid separation, can be used to increase digestate performance, especially those with relatively higher dry matter content (Sørensen and Thomsen, 2005).

The application of digestates with a high fraction of undecomposed organic material, characterized by a high C/N ratio, promotes soil N immobilization. Microbial assimilation of N linked to respiration of manure C typically starts immediately after manure application and reaches a peak after 1-4 weeks, after which mineralization starts (Sørensen, 2004). Immobilization reduces readily available nitrogen to crops, reducing fertilizer value.

With increasing utilization of recalcitrant substrates as inputs, only 40-70% of the biomass is converted into biogas, resulting in a digestate with a significantly high degradable organic matter (Romio et al., 2021). Incorporating such digestates increases oxygen demand in the soil, inducing denitrification. Due to denitrification, loss of N from manure application in the form of N_2 and N_2O is usually high compared to synthetic fertilizers (Velthof et al., 2003). Denitrification losses are significantly influenced by manure distribution, soil water content and soil type (Webb et al., 2013). Assessing the benefits of incorporating manure into AD should include a mass flow approach at the whole farm level, including factors such as NH_3 volatilization, reduced carbon sequestration after digestate application and methane emissions (Chadwick et al., 2011; Møller et al., 2022).

6   Pre- and post-treatment of biomass for anaerobic digestion

During the AD process, hydrolysis and methanogenesis are limiting stages due to the recalcitrant nature of biomass and the formation and accumulation of undesirable intermediary products such as volatile fatty acids (Rajagopal et al., 2013; Siegert and Banks, 2005). These limitations result in less stable biogas yields with digestate characterized by partially-decomposed organic matter with reduced nutrient availability. Ways of improving the AD process include incorporating a pre-treatment step for lignocellulosic biomass before AD or incorporating a post-treatment stage as illustrated in Fig. 3.

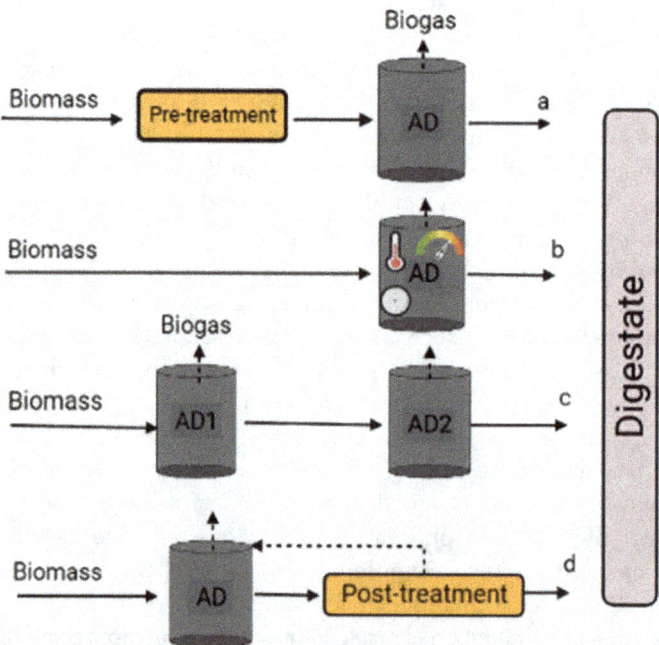

Figure 3 Schematic diagram of how anaerobic digestion process can be optimized for higher biogas yield and enhanced nutrient availability, (a) illustrates the incorporation of a pre-treatment step before AD, (b) optimization of operation conditions, i.e. temperature and HRT, (c) extending the HRT by adjusting reactors configuration to have primary and secondary AD steps and (d) a post-treatment step, i.e. solid-liquid separation.

6.1 Pre-treatment of biomass

Pre-treatment of biowastes facilitates degradation by enzymes and microbes during AD by increasing reaction kinetics and nutrient solubilization (Sarker et al., 2019; Zheng et al., 2014). Pre-treatment also helps to overcome lignin recalcitrance and crystalline cellulose barriers during digestion, optimizing process efficiency and increasing biogas production and digestate quality (Meegoda et al., 2018; Sarker et al., 2019). Several studies have reported positive effects of pre-treatment techniques on biogas yield (Harris and McCabe, 2015; Hernández-Beltrán et al., 2019; Sayara and Sánchez, 2019).

Pre-treatment techniques include physical, chemical and biological methods. Physical pre-treatment techniques e.g. include grinding, milling, extrusion, sonication, high pressure, microwave digestion, ozonolysis and electrokinetic treatments. These increase the accessible surface area of the substrate and decrease the crystallinity of cellulose during AD (Kumari and Singh, 2018; Sayara and Sánchez, 2019). The increase in surface area and release of intracellular components increases the degradation rate and accelerates the AD process, increasing methane yield.

Nyang'au et al. (2023), e.g. found that an electrokinetic pre-treatment step ($P < 0.05$) increased NH_4^+-N/total N in digestates before the second AD step, but the effect levelled off after the secondary digestion step (Fig. 4). Electrokinetic and ultrasonication pre-treatments tended to increase the mineral fertilizer equivalent of ammonium-N in digestates (Table 4). They concluded that these pre-treatment steps could improve digestate fertilizing properties such as lower dry matter content and viscosity alongside increasing biogas yields.

In selecting an appropriate pre-treatment technique to incorporate into the biogas digester, cost and energy consumption must be adequately evaluated (Meegoda et al., 2018). Pre-treatment also needs to significantly improve cellulose and hemicellulose solubilization at low cost (Kim et al., 2014). Other factors include increasing biogas yield, improving digestate fertilizer properties and reducing biowaste treatment time (Nyang'au et al., 2023). High biogas recovery after pre-treatment can reduce labile organic matter in digestates, increasing stability and the ratio of available N.

6.2 Post-treatment of biomass

Due to regulations limiting nitrogen application up to 170 kg N ha^{-1} in many EU countries (EU nitrates directive), digestates often have to be transported to other areas for use. The high water content of digestates increases costs of transportation and storage, potentially exceeding the fertilizer value of the digestate (Plana and Noche, 2016). To deal with these challenges and improve

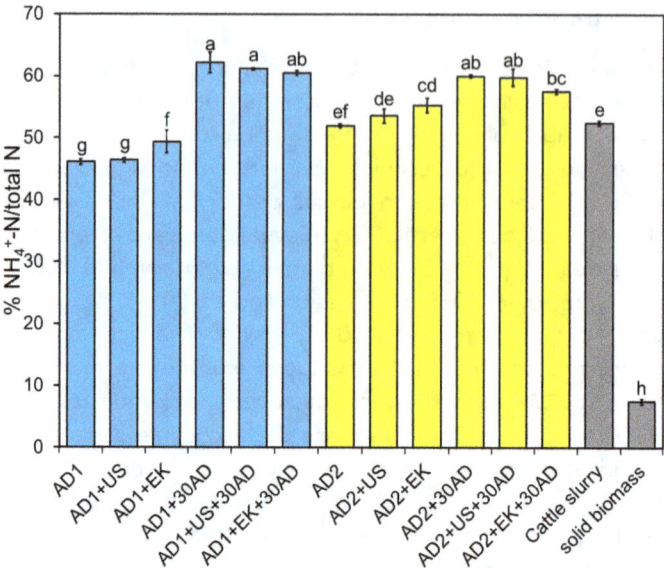

Figure 4 Effect of pre-treatment techniques, electrokinetic (EK) and ultrasonication (US), and two-step-AD (+30AD) on percentage NH_4^+-N/total N in two digestates (AD1 and AD2) sourced from two biogas plants with contrasting HRT in a primary AD step. Letters indicate significant differences among the treatments estimated with Tukey's test ($P <$ 0.05). Error bars indicate standard deviation (n=3). Reprinted from Nyang'au et al. (2023) with permission.

digestate NUE, post-treatment technologies such as solid-liquid separation can be adopted for more tailored field applications.

Separation of digestate results in a liquid fraction with a high proportion of ammonium nitrogen and potassium, and a solid fraction rich in phosphorus and organic matter which is lighter and cheaper to transport. The liquid fraction could replace nitrogen-based mineral fertilizer, while the solid fraction of the digestate can be utilized in regions with P-deficient soils (Fangueiro et al., 2011; Romero-Güiza et al., 2016). The liquid fraction can also be further treated to meet legislation requirements such as RENURE criteria relating to N recovery from manure (Reuland et al., 2021) or by recovering nutrients such as N through ammonia stripping. The low dry matter content of the liquid fraction makes it less viscous and able to infiltrate soil faster, reducing the risk of NH_3 volatilization and improving NUE after surface application (Sørensen and Thomsen, 2005). The liquid fraction also has a low C/N ratio and a high potential N fertilizer value (Webb et al., 2013). Sørensen and Thomsen (2005) observed improved NUE after surface application of separated liquid fraction of slurry in winter wheat, though only techniques including rapid incorporation of the solid fraction and surface banding or injection of the liquid fraction improved overall NUE.

Table 4 Recovery of labelled ammonium-N, mineral fertilizer equivalence (MFE) of ammonium-N in digestates calculated based on ^{15}N uptake in winter wheat and spring barley in above-ground biomass relative to ^{15}N uptake in mineral fertilizer and estimated ammonia losses in winter wheat after surface banding application of digestates pre-treated by electrokinetic (EK) and ultrasonication (US) in a two-step AD

	Secondary AD step	Winter wheat		Spring barley		Estimated % NH$_4^+$-N losses
		% Recovery	% MFE	% Recovery	% MFE	
AD1	–	31.1	51.7 (1.4)[b]	36.0	82.4 (1.0)[a]	30.7[ab]
AD1+EK	–	32.2	53.6 (0.7)[b]	36.4	83.2 (5.9)[a]	29.7[abc]
AD1+US	–	31.6	52.5 (2.4)[b]	37.2	85.1 (4.7)[a]	32.6[a]
AD1[+AD30]	+	37.7	62.6 (1.8)[a]	36.4	88.3 (1.3)[a]	20.7[bcd]
AD1+EK	+	37.6	62.6 (1.3)[a]	35.7	81.5 (3.1)[a]	19.0[cd]
AD1+US	+	38.5	63.9 (0.1)[a]	35.1	80.2 (3.1)[a]	16.3[d]

Means followed by different letters within each column are significantly different ($P < 0.05$). For % MFE, values correspond to averages and standard errors in brackets ($n = 4$). The ammonia loss is estimated from the difference in MFE between spring barley and winter wheat, assuming no ammonia loss by injection to spring barley. Data from Nyang'au et al. (2023).

7   Optimization of anaerobic digestion operations to optimize digestate quality

There are a huge range of variables which affect AD performance in terms of biogas or digestate yield and quality (Sarker et al., 2019; Veluchamy et al., 2019). However, optimization of AD is often based on biogas and methane yield with limited attention given to the digestate quality (Logan and Visvanathan, 2019). This results in a digestate with undesirable properties such as high dry matter content. Extending biomass hydraulic retention time (HRT) using staged reactors e.g. can result in both higher biogas yield and high nutrient mineralization in digestate (Feng et al., 2017).

In Denmark, e.g. there has been an increasing trend of incorporating secondary reactors into biogas plants to both optimize biogas yields and utilization of biomass such as straw (Møller and Nielsen, 2016). Increasing the residence time of substrates in digesters enhances degradation, resulting in more biogas recovery, nutrient solubilization and stabilized digestate. Applying stabilized digestate from longer HRT to soil results in less C mineralization without affecting soil carbon retention (Nyang'au et al., 2022). Nyang'au et al. (2023) found that incorporating a secondary digestion step for biowastes to increase HRT could, on average, increase winter wheat N uptake from 39% to 42% following digestate surface application and from 75% to 86% following digestate injection in soil. Through ^{15}N-labelling of ammonium-N in digestates, they also found that the extra digestion step significantly reduced ammonia losses through volatilization (Table 4).

There are typically two temperature regimes used in AD:

- Mesophilic; and
- Thermophilic.

Mesophilic digestion operates at around 35-45°C and and, whilst requiring less energy, is associated with a slower degradation rate and less biogas production. Thermophilic digestion occurs at higher temperatures (around 50-55°C) and is characterized by faster degradation, potentially increasing biogas yield and nutrient solubilization (Hartmann and Ahring, 2006). Hossain et al. (2021) e.g. found that thermophilic temperatures marginally increased inorganic N release of digestate to soil compared to digestion at mesophilic temperature. Thermophilic conditions could provide additional benefits by destroying pathogens such *Escherichia coli* and *Salmonella* spp., weed seeds and other foreign material before field application (Smith et al., 2005; Sørensen et al., 2019).

8 References

Aarnink, A. and Ogink, N. (2007). Environmental impact of daily removal of pig manure with a conveyer belt system. In: *International Symposium on Air Quality and Waste Management for Agriculture.* Broomfield, CO.

Aarnink, A. J. A., Veld, J. H. I. T. and Hol, A. I. V. (2007). *Kempfarm vleesvarkensstal: milieu emissies en Investeringskosten.* Wageningen University and Research.

Abdalla, C. W. (2002). The industrialization of agriculture; implications for public concern and environmental consequences of intensive livestock operations. *Penn State Environmental Law* 10: 175-191.

Aguirre-Villegas, H. A., Larson, R. and Reinemann, D. J. (2014). From waste-to-worth: Energy, emissions, and nutrient implications of manure processing pathways. *Biofuels, Bioproducts and Biorefining* 8(6): 770-793.

Aguirre-Villegas, H. A., Larson, R. A. and Sharara, M. A. (2019). Anaerobic digestion, solid-liquid separation, and drying of dairy manure: measuring constituents and modeling emission. *Science of the Total Environment* 696: 134059.

Amann, M., Anderl, M., Borken-Kleefeld, J., Cofala, J., Heyes, C., Höglund Isaksson, L., Kiesewetter, G., Klimont, Z., Moosmann, L. and Rafaj, P. (2018). *Progress Towards the Achievement of the EU's Air Quality and Emissions Objectives.* IIASA, Laxenburg, Austria.

Amon, B., Stouman Jensen, L. S., Groenestein, K. and Sutton, M. (2021). Sustainable nitrogen management for housed livestock, manure storage and manure processing. In: Baines, R. (Ed.). *Reducing greenhouse gas emissions from livestock production.* Burleigh Dodds Science Publishing, Cambridge, 46 pages.

Anderson, C., Malambo, D. H., Gonzalez Perez, M. E., Nobela, H. N., de Pooter, L., Spit, J., Hooijmans, C. M., de Vossenberg, Jv., Greya, W., Thole, B., van Lier, J. B. and Brdjanovic, D. (2015). Lactic acid fermentation, urea and lime addition: Promising faecal sludge sanitizing methods for emergency sanitation. *International Journal of Environmental Research and Public Health* 12(11): 13871-13885.

Barampouti, E. M., Mai, S., Malamis, D., Moustakas, K. and Loizidou, M. (2020). Exploring technological alternatives of nutrient recovery from digestate as a secondary resource. *Renewable and Sustainable Energy Reviews* 134: 110379.

Bastami, M. S. B., Jones, D. L. and Chadwick, D. R. (2016). Reduction of methane emission during slurry storage by the addition of effective microorganisms and excessive carbon source from brewing sugar. *Journal of Environmental Quality* 45(6): 2016-2022.

Burney, J. A., Davis, S. J. and Lobell, D. B. (2010). Greenhouse gas mitigation by agricultural intensification. *Proceedings of the National Academy of Sciences of the United States of America* 107(26): 12052-12057.

Cao, L., Keener, H., Huang, Z., Liu, Y., Ruan, R. and Xu, F. (2020). Effects of temperature and inoculation ratio on methane production and nutrient solubility of swine manure anaerobic digestion. *Bioresource Technology* 299: 122552.

Case, S. D. C. and Jensen, L. S. (2019). Nitrogen and phosphorus release from organic wastes and suitability as bio-based fertilizers in a circular economy. *Environmental Technology* 40(6): 701-715.

Chadwick, D., Sommer, S., Thorman, R., Fangueiro, D., Cardenas, L., Amon, B. and Misselbrook, T. (2011). Manure management: implications for greenhouse gas emissions. *Animal Feed Science and Technology* 166-167: 514-531.

Clemens, J., Trimborn, M., Weiland, P. and Amon, B. (2006). Mitigation of greenhouse gas emissions by anaerobic digestion of cattle slurry. *Agriculture, Ecosystems and Environment* 112(2-3): 171-177.

Chojnacka, K., Moustakas, K. and Witek-Krowiak, A. (2020). Bio-based fertilizers: A practical approach towards circular economy. *Bioresource Technology* 295: 122223.

Christensen, M. L., Hjorth, M. and Keiding, K. (2009). Characterization of pig slurry with reference to flocculation and separation. *Water Research* 43(3): 773-783.

De Vries, J. W., Aarnink, A. J. A., Groot Koerkamp, P. W. G. and De Boer, I. J. M. (2013). Life cycle assessment of segregating fattening pig urine and feces compared to conventional liquid manure management. *Environmental Science and Technology* 47(3): 1589-1597.

De Vries, J. W., Groenestein, C. M., Schröder, J. J., Hoogmoed, W. B., Sukkel, W., Groot Koerkamp, P. W. G. and De Boer, I. J. M. (2015). Integrated manure management to reduce environmental impact: II. Environmental impact assessment of strategies. *Agricultural Systems* 138: 88-99.

De Vries, J. W., Vinken, T. M. W. J., Hamelin, L. and De Boer, I. J. M. (2012). Comparing environmental consequences of anaerobic mono- and co-digestion of pig manure to produce bio-energy – A life cycle perspective. *Bioresource Technology* 125: 239-248.

De Vrieze, J., Colica, G., Pintucci, C., Sarli, J., Pedizzi, C., Willeghems, G., Bral, A., Varga, S., Prat, D., Peng, L., Spiller, M., Buysse, J., Colsen, J., Benito, O., Carballa, M. and Vlaeminck, S. E. (2019). Resource recovery from pig manure via an integrated approach: A technical and economic assessment for full-scale applications. *Bioresource Technology* 272: 582-593.

Dennehy, C., Lawlor, P. G., Jiang, Y., Gardiner, G. E., Xie, S., Nghiem, L. D. and Zhan, X. (2017). Greenhouse gas emissions from different pig manure management techniques: A critical analysis. *Frontiers of Environmental Science and Engineering* 11(3): 1-16.

Dittmer, K. M., Darby, H. M., Goeschel, T. R. and Adair, E. C. (2020). Benefits and tradeoffs of reduced tillage and manure application methods in a Zea mays silage system. *Journal of Environmental Quality* 49(5): 1236-1250.

Directive 91/676/EEC of the Council of the European Communities of 12 December 1991 concerning the protection of waters against pollution caused by nitrates from agricultural sources.

Emmerling, C., Krein, A. and Junk, J. (2020). Meta-analysis of strategies to reduce NH3 emissions from slurries in European agriculture and consequences for greenhouse gas emissions. *Agronomy* 10(11): 1633.

Ershadi, S. Z., Dias, G., Heidari, M. D. and Pelletier, N. (2020). Improving nitrogen use efficiency in crop-livestock systems: a review of mitigation technologies and management strategies, and their potential applicability for egg supply chains. *Journal of Cleaner Production* 265: 121671.

Erwiha, G. M., Ham, J., Sukor, A., Wickham, A. and Davis, J. G. (2020). Organic fertilizer source and application method impact ammonia volatilization. *Communications in Soil Science and Plant Analysis* 51(11): 1469-1482.

Eurostat (2022). *Key Figures on the European Food Chain* (2022 edn.). 7 December.

Eurostat (2016). Agri-environmental indicator - mineral fertiliser consumption [WWW Document]. Available at: https://ec.europa.eu/eurostat/statistics-explained/index

.php/Agri-environmental_indicator_-_mineral_fertiliser_consumption (accessed 4 January 2023).

Eurostat (2021). Glossary: Fertiliser. Available at: https://ec.europa.eu/eurostat/statistics -explained/index.php/Glossary:Fertiliser.

Fangueiro, D., Coutinho, J., Chadwick, D., Moreira, N. and Trindade, H. (2008). Effect of cattle slurry separation on greenhouse gas and ammonia emissions during storage. *Journal of Environmental Quality* 37(6): 2322-2331.

Fangueiro, D., Ribeirom,H., Vasconcelos, E., Coutinho, J. and Cabral, F. (2009). Treatment by acidification followed by solid-liquid separation affects slurry and slurry fractions composition and their potential of N mineralization. *Bioresource Technology* 100(20): 4914-4917.

Fangueiro, D., Surgy, S., Napier, V., Menaia, J., Vasconcelos, E. and Coutinho, J. (2014). Impact of slurry management strategies on potential leaching of nutrients and pathogens in a sandy soil amended with cattle slurry. *Journal of Environmental Management* 146: 198-205.

Fangueiro, D., Hjorth, M. and Gioelli, F. (2015). Acidification of animal slurry-a review. *Journal of Environmental Management* 149: 46-56.

Fangueiro, D., Merino, P., Pantelopoulos, A., Pereira, J. L. S., Amon, B. and Chadwick, D. R. (2023). The implications of animal manure management on ammonia and greenhouse gas emissions. In: Bartzanas, T. (Ed.). *Technology for Environmentally Friendly Livestock Production Smart Animal Production*. Springer, Cham. https://doi .org/10.1007/978-3-031-19730-7_5.

Feng, L., Wahid, R., Ward, A. J. and Møller, H. B. (2017). Anaerobic co-digestion of cattle manure and meadow grass: effect of serial configurations of continuous stirred tank reactors (CSTRs). *Biosystems Engineering* 160: 1-11.

Foged, H., Flotats, X., Blasi, A. B., Palatsi, J., Magri, A. and Schelde, K. M. (2011). Inventory of manure processing activities in Europe. Technical Report No. I concerning 'Manure Processing Activities in Europe' to the European Commission, Directorate-General Environment.

Fuchs, A., Dalby, F. R., Liu, D., Kai, P. and Feilberg, A. (2021). Improved effect of manure acidification technology for gas emission mitigation by substituting sulfuric acid with acetic acid. *Cleaner Engineering and Technology* 4: 100263.

Gioelli, F., Grella, M., Scarpeci, T. E., Rollè, L., Pierre, F. D. and Dinuccio, E. (2022). Bio-acidification of cattle slurry with whey reduces gaseous emission during storage with positive effects on biogas production. *Sustainability* 14(19): 12331.

Hafner, S. D., Pacholski, A., Bittman, S., Burchill, W., Bussink, W., Chantigny, M., Carozzi, M., Génermont, S., Häni, C., Hansen, M. N., Huijsmans, J., Hunt, D., Kupper, T., Lanigan, G., Loubet, B., Misselbrook, T., Meisinger, J. J., Neftel, A., Nyord, T., Pedersen, S. V., Sintermann, J., Thompson, R. B., Vermeulen, B., Vestergaard, A. V., Voylokov, P., Williams, J. R. and Sommer, S. G. (2018). The ALFAM2 database on ammonia emission from field-applied manure: description and illustrative analysis. *Agricultural and Forest Meteorology* 258: 66-79.

Hagos, K., Zong, J., Li, D., Liu, C. and Lu, X. (2017). Anaerobic co-digestion process for biogas production: progress, challenges and perspectives. *Renewable and Sustainable Energy Reviews* 76: 1485-1496.

Harris, P. W. and McCabe, B. K. (2015). Review of pre-treatments used in anaerobic digestion and their potential application in high-fat cattle slaughterhouse wastewater. *Applied Energy* 155: 560-575.

Hartmann, H. and Ahring, B. K. (2006). Strategies for the anaerobic digestion of the organic fraction of municipal solid waste: An overview. *Water Science and Technology* 53(8): 7–22.

Hernández-Beltrán, J. U., Hernández-De Lira, I. O., Cruz-Santos, M. M., Saucedo-Luevanos, A., Hernández-Terán, F. and Balagurusamy, N. (2019). Insight into pretreatment methods of lignocellulosic biomass to increase biogas yield: Current state, challenges, and opportunities. *Applied Sciences* 9(18): 3721.

Hjorth, M., Christensen, K. V., Christensen, M. L. and Sommer, S. G. (2010). Solid-liquid separation of animal slurry in theory and practice. A review. *Agronomy for Sustainable Development* 30(1): 153–180.

Holly, M. A., Larson, R. A., Powell, J. M., Ruark, M. D. and Aguirre-Villegas, H. (2017). Greenhouse gas and ammonia emissions from digested and separated dairy manure during storage and after land application. *Agriculture, Ecosystems and Environment* 239: 410–419.

Hossain, M. K., Islam, M. R., Jahiruddin, M., Sorensen, P., Møller, H. B. and Islam, M. S. (2021). Effect of anaerobic digestion temperature and manure type on N and S mineralization. *Communications in Soil Science and Plant Analysis* 52(20): 2431–2444.

Hou, Y., Velthof, G. L., Case, S. D. C., Oelofse, M., Grignani, C., Balsari, P., Zavattaro, L., Gioelli, F., Bernal, M. P., Fangueiro, D., Trindade, H., Jensen, L. S. and Oenema, O. (2018). Stakeholder perceptions of manure treatment technologies in Denmark, Italy, the Netherlands and Spain. *Journal of Cleaner Production* 172: 1620–1630.

Hou, Y., Velthof, G. L., Lesschen, J. P., Staritsky, I. G. and Oenema, O. (2017). Nutrient recovery and emissions of ammonia, nitrous oxide, and methane from animal manure in Europe: effects of manure treatment technologies. *Environmental Science and Technology* 51(1): 375–383.

Hou, Y., Velthof, G. L. and Oenema, O. (2015). Mitigation of ammonia, nitrous oxide and methane emissions from manure management chains: A meta-analysis and integrated assessment. *Global Change Biology* 21(3): 1293–1312.

Kavanagh, I., Fenton, O., Healy, M. G., Burchill, W., Lanigan, G. J. and Krol, D. J. (2021). Mitigating ammonia and greenhouse gas emissions from stored cattle slurry using agricultural waste, commercially available products and a chemical acidifier. *Journal of Cleaner Production* 294: 126251.

Keskinen, R., Termonen, M., Salo, T., Luostarinen, S. and Räty, M. (2022). Slurry acidification outperformed injection as an ammonia emission-reducing technique in boreal grass cultivation. *Nutrient Cycling in Agroecosystems* 122(2): 139–156.

Kim, T. H., Oh, K. K., Ryu, H. J., Lee, K. and Kim, T. H. (2014). Hydrolysis of hemicellulose from barley straw and enhanced enzymatic saccharification of cellulose using acidified zinc chloride. *Renewable Energy* 65: 56–63.

Kleinman, P. J. A., Spiegal, S., Liu, J., Holly, M., Church, C. and Ramirez-Avila, J. (2020). Managing animal manure to minimize phosphorus losses from land to water. In: Waldrip, H. M., Pagliari, P. H. and He, Z. *Animal Manure*, pp. 201–228.

Koger, J. B., O'Brien, B. K., Burnette, R. P., Kai, P., van Kempen, M. H. J. G., van Heugten, E. and van Kempen, T. A. T. G. (2014). Manure belts for harvesting urine and feces separately and improving air quality in swine facilities. *Livestock Science* 162: 214–222.

Köninger, J., Lugato, E., Panagos, P., Kochupillai, M., Orgiazzi, A. and Briones, M. J. I. (2021). Manure management and soil biodiversity: towards more sustainable food systems in the EU. *Agricultural Systems* 194: 103251.

Kumari, D. and Singh, R. (2018). Pretreatment of lignocellulosic wastes for biofuel production: A critical review. *Renewable and Sustainable Energy Reviews* 90: 877-891.

Kupper, T., Häni, C., Neftel, A., Kincaid, C., Bühler, M., Amon, B. and VanderZaag, A. (2020). Ammonia and greenhouse gas emissions from slurry storage - a review. *Agriculture, Ecosystems and Environment* 300: 106963.

Li, B., Dinkler, K., Zhao, N., Sobhi, M., Merkle, W., Liu, S., Dong, R., Oechsner, H. and Guo, J. (2020). Influence of anaerobic digestion on the labile phosphorus in pig, chicken, and dairy manure. *Science of the Total Environment* 737: 140234.

Logan, M. and Visvanathan, C. (2019). Management strategies for anaerobic digestate of organic fraction of municipal solid waste: current status and future prospects. *Waste Management and Research* 37(1_suppl): 27-39.

Loussouarn, A., Lagadec, S., Robin, P. and Hassouna, M. (2014). Raclage en «V»: Bilan environnemental et zootechnique lors de sept années de fonctionnement à Guernévez. In: *46èmes Journées de la Recherche Porcine*. Paris, France, pp. 199-204.

Maldaner, L., Wagner-Riddle, C., VanderZaag, A. C., Gordon, R. and Duke, C. (2018). Methane emissions from storage of digestate at a dairy manure biogas facility. *Agricultural and Forest Meteorology* 258: 96-107.

Malique, F., Wangari, E., Andrade-Linares, D. R., Schloter, M., Wolf, B., Dannenmann, M., Schulz, S. and Butterbach-Bahl, K. (2021). Effects of slurry acidification on soil N2O fluxes and denitrification. *Journal of Plant Nutrition and Soil Science* 184(6): 696-708.

Meegoda, J. N., Li, B., Patel, K. and Wang, L. B. (2018). A review of the processes, parameters, and optimization of anaerobic digestion. *International Journal of Environmental Research and Public Health* 15(10): 2224.

Møller, H. B. and Nielsen, K. J. (2016). Biogas taskforce: udvikling og effektivisering af biogasproduktionen i Danmark. DCA-Nationalt Center for Fødevarer og Jordbrug.

Møller, H. B., Sørensen, P., Olesen, J. E., Petersen, S. O., Nyord, T. and Sommer, S. G. (2022). Agricultural biogas production–climate and environmental impacts. *Sustainability* 14(3): 1849.

Moset, V., Fontaine, D. and Møller, H. B. (2017). Co-digestion of cattle manure and grass harvested with different technologies. Effect on methane yield, digestate composition and energy balance. *Energy* 141: 451–460.

Moset, V., Cerisuelo, A., Sutaryo, S. and Møller, H. B. (2012). Process performance of anaerobic co-digestion of raw and acidified pig slurry. *Water Research* 46(16): 5019-5027.

Moset, V., Ottosen, L. D. M., Xavier, C. and Møller, H. B. (2016). Anaerobic digestion of sulfate-acidified cattle slurry: one-stage vs. two-stage. *Journal of Environmental Management* 173: 127-133.

Moyo, L. B., Simate, G. S. and Mutsatsa, T. (2022). Biological acidification of pig manure using banana peel waste to improve the dissolution of particulate phosphorus: A critical step for maximum phosphorus recovery as struvite. *Heliyon* 8(8): e10091.

Niles, M. T., Wiltshire, S., Lombard, J., Branan, M., Vuolo, M., Chintala, R. and Tricarico, J. (2022). Manure management strategies are interconnected with complexity across US dairy farms. *PLoS ONE* 17(6): e0267731.

Niles, M. T., Horner, C., Chintala, R. and Tricarico, J. (2019). A review of determinants for dairy farmer decision making on manure management strategies in high-income countries. *Environmental Research Letters* 14(5): 053004.

Nyang'au, J. O., Moller, H. B. and Sorensen, P. (2022). Nitrogen dynamics and carbon sequestration in soil following application of digestates from one- and two-step anaerobic digestion. *Science of the Total Environment* 851(1): 158177.

Nyang'au, J. O., Møller, H. B. and Sørensen, P. (2023). Effects of electrokinetic and ultrasonication pre-treatment and two-step anaerobic digestion of biowastes on the nitrogen fertiliser value by injection or surface banding to cereal crops. *Journal of Environmental Management* 326(A): 116699.

Nyord, T., Liu, D., Eriksen, J. and Adamsen, A. P. S. (2013). Effect of acidification and soil injection of animal slurry on ammonia and odour emission. *15th RAMIRAN Conference*, Versailles, France.

Oenema, O. and Tamminga, S. (2005). Nitrogen in global animal production and management options for improving nitrogen use efficiency. *Science in China Series C Life Sciences* 48(S2): 871–887.

Olesen, J. E., Møller, H. B., Petersen, S. O., Sørensen, P., Nyord, T. and Sommer, S. G. (2021). Sustainable biogas-climate and environmental effects of biogas production. DCA Report. Aarhus Universitet, DCA – Danish Centre for Food and Agriculture, p. 85.

Overmeyer, V., Holtkamp, F., Clemens, J., Büscher, W. and Trimborn, M. (2020). Dynamics of different buffer systems in slurries based on time and temperature of storage and their visualization by a new mathematical tool. *Animals: An Open Access Journal from MDPI* 10(4): 724.

Overmeyer, V., Kube, A., Clemens, J., Büscher, W. and Trimborn, M. (2021). One-time acidification of slurry: what is the most effective acid and treatment strategy? *Agronomy* 11(7): 1319.

Pedersen, I. F., Rubæk, G. H. and Sørensen, P. (2017). Cattle slurry acidification and application method can improve initial phosphorus availability for maize. *Plant and Soil* 414(1–2): 143–158.

Pedersen, J., Andersson, K., Feilberg, A., Delin, S., Hafner, S. and Nyord, T. (2021). Effect of exposed surface area on ammonia emissions from untreated, separated, and digested cattle manure. *Biosystems Engineering* 202: 66–78.

Plana, P. V. and Noche, B. (2016). A review of the current digestate distribution models: Storage and transport. *WIT Transactions on Ecology and the Environment* 202: 345–357.

Prado, J., Chieppe, J., Raymundo, A. and Fangueiro, D. (2020). Bio-acidification and enhanced crusting as an alternative to sulphuric acid addition to slurry to mitigate ammonia and greenhouse gases emissions during short term storage. *Journal of Cleaner Production* 263: 121443.

Prado, J., Ribeiro, H., Alvarenga, P. and Fangueiro, D. (2022). A step towards the production of manure-based fertilizers: disclosing the effects of animal species and slurry treatment on their nutrients content and availability. *Journal of Cleaner Production* 337: 130369.

Rajagopal, R., Massé, D. I. and Singh, G. (2013). A critical review on inhibition of anaerobic digestion process by excess ammonia. *Bioresource Technology* 143: 632–641.

Regueiro, I., Coutinho, J., Balsari, P., Popovic, O. and Fangueiro, D. (2016). Acidification of pig slurry before separation to improve slurry management on farms. *Environmental Technology* 37(15): 1906–1913.

Regueiro, I., Coutinho, J. and Fangueiro, D. (2016). Alternatives to sulfuric acid for slurry acidification: impact on slurry composition and ammonia emissions during storage. *Journal of Cleaner Production* 131: 296–307.

Regueiro, I., Siebert, P., Liu, J., Müller-Stöver, D. and Jensen, L. S. (2020). Acidified animal manure products combined with a nitrification inhibitor can serve as a starter fertilizer for maize. *Agronomy* 10(12): 1941.

Regueiro, I., Gómez-Muñoz, B., Lübeck, M., Hjorth, M. and Jensen, L. S. (2022). Bio-acidification of animal slurry: efficiency, stability and the mechanisms involved. *Bioresource Technology Reports* 19: 101135.

Regulation 1069/2009 of the European Parliament and of the Council of 21 October 2009 laying down health rules as regards animal by-products and derived products not intended for human consumption and repealing Regulation (EC) No 1774/2002 (Animal by-products Regulation).

Reuland, G., Sigurnjak, I., Dekker, H., Michels, E. and Meers, E. (2021). The potential of digestate and the liquid fraction of digestate as chemical fertiliser substitutes under the RENURE criteria. *Agronomy* 11(7): 1374.

Rodrigues, J., Alvarenga, P., Silva, A. C., Brito, L., Tavares, J. and Fangueiro, D. (2021). Animal slurry sanitization through PH adjustment: process optimization and impact on slurry characteristics. *Agronomy* 11(3): 517.

Romero-Güiza, M. S., Mata-Alvarez, J., Chimenos Rivera, J. M. and Astals Garcia, S. (2016). Nutrient recovery technologies for anaerobic digestion systems: an overview. *Revista ION* 29(1): 7–26.

Romio, C., Kofoed, M. V. W.and Moller, H. B. (2021). Digestate post-treatment strategies for additional biogas recovery: a review. *Sustainability* 13(16): 9295.

Sarker, S., Lamb, J. J., Hjelme, D. R. and Lien, K. M. (2019). A review of the role of critical parameters in the design and operation of biogas production plants. *Applied Sciences* 9(9): 1915.

Sayara, T. and Sánchez, A. (2019). A review on anaerobic digestion of lignocellulosic wastes: pretreatments and operational conditions. *Applied Sciences* 9(21): 4655.

Schreiber, M., Bazaios, E., Ströbel, B., Wolf, B., Ostler, U., Gasche, R., Schlingmann, M., Kiese, R. and Dannenmann, M. (2023). Impacts of slurry acidification and injection on fertilizer nitrogen fates in grassland. *Nutrient Cycling in Agroecosystems* 125(2): 171–186.

Sepperer, T., Petutschnigg, A. and Steiner, K. (2021). Effect of flushing milk and acidic whey on pH and nitrogen loss of cattle manure slurry. *Atmosphere* 12(9): 1222.

Siegert, I. and Banks, C. (2005). The effect of volatile fatty acid additions on the anaerobic digestion of cellulose and glucose in batch reactors. *Process Biochemistry* 40(11): 3412–3418.

Smith, S. R., Lang, N. L., Cheung, K. H. M. and Spanoudaki, K. (2005). Factors controlling pathogen destruction during anaerobic digestion of biowastes. *Waste Management* 25(4): 417–425.

Søgaard, H. T., Sommer, S. G., Hutchings, N. J., Huijsmans, J. F. M., Bussink, D. W. and Nicholson, F. (2002). Ammonia volatilization from field-applied animal slurry—the ALFAM model. *Atmospheric Environment* 36(20): 3309–3319.

Sommer, S. G. and Husted, S. (1995). A simple model of pH in slurry. *The Journal of Agricultural Science* 124(3): 447–453.

Sørensen, C. G., Jacobsen, B. H. and Sommer, S. G. (2003). An assessment tool applied to manure management systems using innovative technologies. *Biosystems Engineering* 86(3): 315–325. https://doi.org/10.1016/S1537-5110(03)00137-5.

Sørensen, P. (2004). Immobilisation, remineralisation and residual effects in subsequent crops of dairy cattle slurry nitrogen compared to mineral fertiliser nitrogen. *Plant and Soil* 267(1-2): 285-296.

Sørensen, P., Bechini, L. and Jensen, L. S. (2019). *Manure management in organic farming.* In: Köpke, U. (ed.) *Improving organic crop cultivation,* Burleigh Dodds Science Publishing: Cambridge, UK.

Sørensen, P., Khan, A. R., Møller, H. B. and Thomsen, I. K. (2012). Effects of anaerobic digestion of organic manures on N turnover and N utilization. *Proceedings of the 17th Nitrogen Workshop-Innovations for Sustainable Use of Nitrogen Resources*, pp. 80-81.

Sørensen, P. and Thomsen, I. K. (2005). Separation of pig slurry and plant utilization and loss of nitrogen-15-labeled slurry nitrogen. *Soil Science Society of America Journal* 69(5): 1644-1651.

Sommer, S. G. and Husted, S. (1995). The chemical buffer system in raw and digested animal slurry. *The Journal of Agricultural Science* 124(1): 45-53.

Sutaryo, S., Ward, A. J. and Møller, H. B. (2013). Anaerobic digestion of acidified slurry fractions derived from different solid-liquid separation methods. *Bioresource Technology* 130: 495-501.

Ten Hoeve, M., Gómez-Muñoz, B., Jensen, L. S. and Bruun, S. (2016). Environmental impacts of combining pig slurry acidification and separation under different regulatory regimes - A life cycle assessment. *Journal of Environmental Management* 181: 710-720.

Thangarajan, R., Bolan, N. S., Tian, G., Naidu, R. and Kunhikrishnan, A. (2013). Role of organic amendment application on greenhouse gas emission from soil. *Science of the Total Environment* 465: 72-96.

Varma, V. S., Parajuli, R., Scott, E., Canter, T., Lim, T. T., Popp, J. and Thoma, G. (2021). Dairy and swine manure management-Challenges and perspectives for sustainable treatment technology. *Science of the Total Environment* 778: 146319.

Velthof, G. L., Kuikman, P. J. and Oenema, O. (2003). Nitrous oxide emission from animal manures applied to soil under controlled conditions. *Biology and Fertility of Soils* 37(4): 221-230.

Veluchamy, C., Gilroyed, B. H. and Kalamdhad, A. S. (2019). Process performance and biogas production optimizing of mesophilic plug flow anaerobic digestion of corn silage. *Fuel* 253: 1097-1103.

Vu, P. T., Melse, R. W., Zeeman, G. and Groot Koerkamp, P. W. G. (2016). Composition and biogas yield of a novel source segregation system for pig excreta. *Biosystems Engineering* 145: 29-38.

Wagner, C., Nyord, T., Vestergaard, A. V., Hafner, S. D. and Pacholski, A. S. (2021). Acidification effects on in situ ammonia emissions and cereal yields depending on slurry type and application method. *Agriculture* 11(11): 1053.

Webb, J., Pain, B., Bittman, S. and Morgan, J. (2010). The impacts of manure application methods on emissions of ammonia, nitrous oxide and on crop response–a review. *Agriculture, Ecosystems and Environment* 137(1-2): 39-46.

Webb, J., Sørensen, P., Velthof, G., Amon, B., Pinto, M., Rodhe, L., Salomon, E., Hutchings, N., Burczyk, P. and Reid, J. (2013). An assessment of the variation of manure nitrogen efficiency throughout Europe and an appraisal of means to increase manure-N efficiency. *Advances in Agronomy*: 371-442.

Welch, R. M. (2002). The impact of mineral nutrients in food crops on global human health. *Plant and Soil* 247(1): 83–90.

Wright, I. A., Tarawali, S., Blummel, M., Gerard, B., Teufel, N. and Herrero, M. (2012). Integrating crops and livestock in subtropical agricultural systems. *Journal of the Science of Food and Agriculture* 92(5): 1010–1015.

Yuan, Z., Pan, X., Chen, T., Liu, X., Zhang, Y., Jiang, S., Sheng, H. and Zhang, L. (2018). Evaluating environmental impacts of pig slurry treatment technologies with a life-cycle perspective. *Journal of Cleaner Production* 188: 840–850.

Zhang, X., Fang, Q., Zhang, T., Ma, W., Velthof, G. L., Hou, Y., Oenema, O. and Zhang, F. (2020). Benefits and trade-offs of replacing synthetic fertilizers by animal manures in crop production in China: A meta-analysis. *Global Change Biology* 26(2): 888–900.

Zheng, Y., Zhao, J., Xu, F. Q. and Li, Y. (2014). Pretreatment of lignocellulosic biomass for enhanced biogas production. *Progress in Energy and Combustion Science* 42: 35–53.

Chapter 4

Optimizing livestock manure as a biofertilizer and bioenergy source

V. Riau, L. Morey, R. Cáceres, M. Cerrillo and A. Bonmatí, Institute of Agrifood Research and Technology (IRTA), Spain; and A. Robles, BETA Tech Center (UVIC-UCC), Spain

1 Introduction

2 Anaerobic digestion

3 Mechanical separation

4 Composting

5 Struvite precipitation

6 Stripping/scrubbing

7 Membrane filtration

8 Bioelectrochemical systems

9 Case study: farm for the future

10 Conclusion and future trends

11 Where to look for further information

12 References

1 Introduction

In 2020, 11.2 million tonnes of mineral fertilizer was used in the European agricultural sector. Ten million tonnes was mineral nitrogen (N) and 1.2 million tonnes was mineral phosphorus (P) (Eurostat, 2022). More than 45% of the total N applied on agricultural fields comes from mineral fertilizers (Hendriks et al., 2022).

Energy from natural gas is typically used to manufacture nitrogenous fertilizers which means its cost is highly dependent on oil prices. The N-based fertilizer industry in Europe depends to a large extent on gas from Russia, which also plays a key role in the global production of P rock-based fertilizers. Moreover, phosphorus extraction is mainly performed outside the EU, leading to high production and transportation costs, which are also related to oil prices. The current geopolitical situation is thus leading to an increase in the price of N and P mineral fertilizer that will have a direct impact on the use of fertilizers in agriculture in the EU (Eurostat, 2022). The current worldwide energy crisis

http://dx.doi.org/10.19103/AS.2023.0120.11

and dependence on non-renewable energy resources has therefore driven interest in the production of biobased fertilizers to secure a more sustainable and resilient agricultural sector (Hendriks et al., 2022).

In this context, livestock manure should be considered as a resource rather than waste as it has an abundance of available nutrients for crops, carbon for bioenergy production and can promote soil carbon stocks. Global livestock production in 2018 was around 1450 million heads, comprising cattle (965 million), pigs (242 million) and chickens (237 million), generating approximately 55 billion tonnes of manure. In terms of nutrient content, this represents around 125 million tonnes of N, only 27% of which is processed in some way. An estimated 12–14 million tonnes per year of P are estimated to be excreted by livestock worldwide (Nagarajan et al., 2023).

Circular agriculture focuses on using minimal amounts of external inputs, closing nutrient loops, regenerating soils and minimizing impacts on the environment, thus aiding the transition towards sustainable and resilient energy and farming systems (Kristinn et al., 2021). Using manure as a source of macro- and micronutrients in agriculture potentially enables crop and livestock production that does not deplete non-renewable sources and does not harm the environment, since it reduces dependence on mineral fertilizers (Prado et al., 2022). In Europe, it is estimated that a circular approach to food systems could reduce the use of chemical fertilizers by 80% (Kristinn et al., 2021).

However, if not managed properly, manure application to soil can lead to detrimental environmental effects from ammonia emissions, pollution of surface water by run-off and groundwater via NO_3^- (nitrate) leaching. Consequently, manure and by-product application to soil in the EU is strictly regulated (maximum 170 kg N ha^{-1} $year^{-1}$ in Nitrate Vulnerable Zones), due to the environmental risks associated with nitrate pollution (European Commission, 1991). To overcome all these limitations, the European Parliament and the EU Council approved the new Fertilizing Product Regulation 1009/2019 (effective as of 16 July 2022), which facilitated access to the EU Single Market for biobased fertilizers. These biobased fertilizers can be awarded the 'CE mark', making it easier to commercialize and, consequently, promote production (Rizzioli et al., 2023). In addition, the possible application of the RENURE criteria (REcovered Nitrogen from manURE) aims to diminish the legal limitations on the application of livestock manure in agriculture. Under these criteria, the application of manure-based fertilizers with a ratio $N_{mineral}/N_{total} > 90\%$, or a ratio of total organic carbon/$N_{total} \leq 3$ would be less restrictive, and the production and use of biobased fertilizers would be promoted (Prado et al., 2022; Hendriks et al., 2022; Brienza et al., 2021).

However, many farmers still prefer the use of mineral fertilizers. This is not only due to logistic/regulative barriers and costs in using alternatives, but also because they are sometimes reluctant to use manure-derived materials that have a wide range of physical and chemical characteristics, and with unbalanced and usually unknown nutrient content. In addition, the investment cost and lack of incentives often makes it difficult to install valorization technologies to process manure at farm level (Khoshnevisan et al., 2021; Tur-Cardona et al., 2018). Therefore, further research is required to establish a systematic framework based on regional requirements to develop an integrated and circular manure nutrient management plan with minimum environmental risks, maximum profit and broad acceptance. A promising solution for the future sustainable development of the agri-food sector is livestock manure management using a biorefinery approach to produce manure-derived fertilizers and/or bioenergy (Awasthi et al., 2022). Technology selection would depend on environmental policies, the investment costs and incentives provided, and the requirement/ market of an end-product in a particular country/city (Khoshnevisan et al., 2021).

This chapter provides an overview of the main technologies for manure valorization into bioenergy and biofertilizer products. Some technologies are already well developed and widely implemented, such as solid-liquid (S/L) separation, anaerobic digestion and composting. Others are more innovative and less developed, including struvite precipitation, ammonia stripping/ scrubbing as well as bioelectrochemical and membrane systems. These technologies can be classified as

- biological (composting, anaerobic digestion and bioelectrochemical systems);
- mechanical (S/L separation); and
- physico-chemical (struvite precipitation, membrane filtration and stripping/scrubbing).

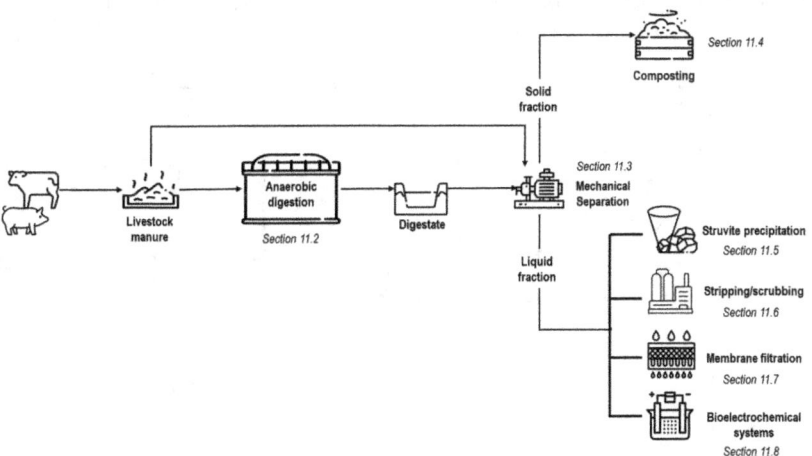

Figure 1 Overview of the different technologies for the production of biobased fertilizers and bioenergy included in this chapter.

However, this chapter has been structured following the scheme shown in Fig. 1, describing the different technologies in the order they are usually implemented along the treatment chain.

2 Anaerobic digestion

Anaerobic digestion (AD) is a well-established and widely implemented technology in the European agricultural sector for treating livestock manure since it generates a renewable fuel in the form of biogas, improves manure fertilizer quality, and reduces odours, pathogens as well as greenhouse gas emissions (Awasthi et al., 2022; Torrellas et al., 2018). The biochemical steps and the factors affecting the AD process have been extensively studied over the last few decades and are well-documented for innumerable feedstocks. This section is therefore mainly focused on the optimization of manure-based bioenergy and biofertilizer production in full-scale AD biogas plants. Figure 2 shows an overview of the main inputs and outputs of the biogas and biomethane production process in the livestock sector.

According to the European Biogas Association, there were almost 15,000 agro-industrial biogas plants in the European Union by the end of 2019 (https://www.europeanbiogas.eu/). However, the AD of manure is only 13-65% efficient when converting organic matter into biogas (Nagarajan et al., 2023; Khoshnevisan et al., 2021), mainly due to the high lignocellulosic content of manure and its low C/N ratio, which results in low methane yields. It has been reported that a suitable C/N ratio from AD systems ranges from 20 to 30, with an optimal of 25 (Zhu et al., 2021). Moreover, biogas yield is strongly dependent on the type of livestock manure and is highly variable among the different studies carried out, ranging from 107 to 438 LCH_4/kg $VS_{removed}$ (Bumharter et al., 2023; Sawatdeenarunat et al., 2016).

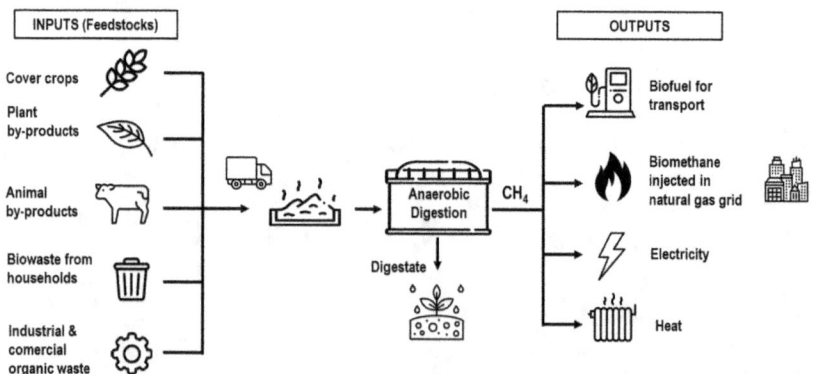

Figure 2 Schematic overview of inputs and outputs of the biogas and biomethane production from livestock manure (modified from EBA's statistical report 2021).

In order to increase bioenergy production from manure, pre-treatment methods such as thermochemical, microwave, ultrasonication and mechanical techniques can be applied to promote the breakdown of the recalcitrant structure of lignocellulosic material and improve the hydrolysis step (Khoshnevisan et al., 2021). However, anaerobic co-digestion of manure, together with other organic substrates containing a high labile carbon content, is the main approach used by agro-industrial biogas plants to improve process efficiency and biogas production (Zhu et al., 2021). It offers multiple advantages compared with single digestion, i.e. better nutrient balance, improved C:N ratio and a robust microbial consortia. Many agricultural feedstocks have been investigated as potential co-substrates over the last few years, including catch crops (Riau et al., 2021), sugar beets (Aboudi et al., 2016), maize silage (Varol and Ugurlu, 2017) or rice straw (Tian et al., 2023). Khoshnevisan et al. (2021) recently published a complete list of synergistic substrates for the anaerobic co-digestion of different livestock manures.

The biogas produced via the AD process has been used in a wide range of applications. To date, the most common industrial-scale application has been heat and electricity generation using a combined heat and power engine, with electrical and thermal conversion efficiencies around 40% and 50%, respectively (Horváth et al., 2016; Sawatdeenarunat et al., 2016). A recent trend is biogas upgrading to remove undesirable compounds, such as H_2S, CO_2, siloxanes, ammonia, moisture and particles from biogas. This can produce biomethane with a CH_4 content greater than 95% (Słupek et al., 2020). The European Biogas Association reported the number of biomethane plants increased from 483 plants in 2018 to 729 units in 2020 (https://www.europeanbiogas.eu/). Biomethane can be used as a fuel for transport, in the form of compressed natural gas, liquid natural gas or injected into a natural gas grid (Fig. 2). Several technologies for biogas purification have been developed in recent years, including adsorption, refrigeration with condensation, membrane technologies and biological methods (Budzianowski, 2016). Apart from its use as a renewable energy resource for heat and electricity generation, biogas can also be subjected to biological and thermochemical processes to produce, for instance, biohydrogen or biomethanol (Sawatdeenarunat et al., 2016).

Apart from biogas production, thermochemical manure treatments such as gasification to produce syngas (H_2 + CO), combustion to generate heat and electricity, or hydrothermal liquefaction/carbonization that can directly convert wet manure into bio-oil or hydrochar, have been developed to obtain manure-based bioenergy. However, the implementation of these technologies at farm level is still very limited. Drawbacks that are yet to be overcome include feedstock composition, undesirable compounds in syngas requiring expensive post-treatment or the configuration of the conversion system, among others (Khoshnevisan et al., 2021). Therefore, further studies and thorough knowledge of these processes in terms of operating conditions are needed to optimize the thermal conversion of livestock manure.

Moreover, in addition to biogas production, the AD of manure also generates a digestate product with good fertilizing value and similar behaviour to mineral fertilizers, since most of the nitrogen, phosphorus, potassium and calcium contained in the input substrates remain in the digestate. The remaining organic carbon also helps to maintain or even increase soil organic matter, which is particularly valuable in marginal or degraded soils (Ehmann et al., 2018). Hence, using anaerobic digestates as organic fertilizer is a common practice that partially reduces the costs derived from both mineral fertilizer and digestate disposal (Kovačić et al., 2022).

Digestate is usually separated into a solid and a liquid fraction in order to reduce the water content and improve transportability, thus reducing management costs (see Section 3). The solid fraction is usually subjected to a composting process to obtain a more stable product and transform (at least partially) N-ammonium into N-nitrate (Cáceres et al., 2015) (see Section 4). The SF can also be valorized into syngas, bio-oil, bio-hydrogen, methanol or hydrochar (Khoshnevisan et al., 2021). Treatment of the liquid fraction has generated considerable interest over the last few years as a feedstock for biobased inorganic fertilizer production such as struvite (Cerrillo et al., 2015) and ammonium sulphate (Sigurnjak et al., 2019) (see Sections 5 and 6, respectively). However, the existing digestate treatments still show considerable optimization potential to increase the effectiveness of nutrient recovery. In this sense, novel technologies to recover nutrients and increase biogas production from digestate, such as membrane systems and bioelectrochemical , have also aroused interest over the last few years (Cerrillo et al., 2023, 2021) (see Sections 7 and 8).

By combining AD and nutrient recovery technologies, energy, nutrients and carbon recovery can be achieved in a more sustainable way. Hence, the integration of these technologies within a single biorefinery is a promising approach that leads to more environmental and economic benefits: reducing nutrient spreading, connecting livestock farmers with industrial sectors, and considering livestock manure as a valuable feedstock for the biobased economy (Khoshnevisan et al., 2021; Sawatdeenarunat et al., 2016).

Livestock manure can be a key feedstock for a lignocellulosic biorefinery, in which materials and chemicals with lignin as a by-product are produced from the cellulose/hemicellulose portion of lignocellulosic biomass (Takkellapati et al., 2018). Hence, Fig. 2 can be transformed into Fig. 3 in order to consider manure AD as a key process within a biorefinery.

The feasibility of AD as a key technology to move to more resilient and sustainable livestock systems has been demonstrated, not only due to the production of biogas as a precursor of environmentally friendly and renewable

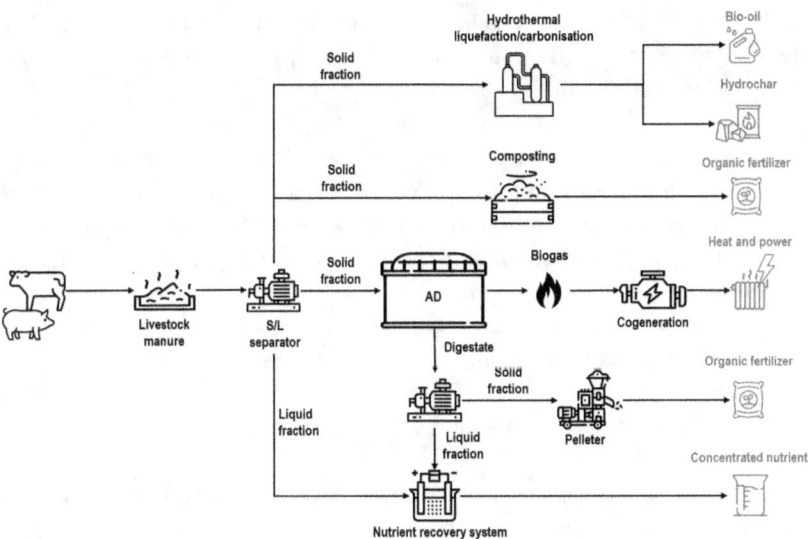

Figure 3 Example of a livestock manure-based biorefinery. Source: modified from Khoshnevisan et al. (2021).

bioenergy and bio-products, but also by generating a valuable fertilizer product that can be further treated to increase nutrient recovery and carbon sequestration in soils.

3 Mechanical separation

Solid/liquid separation is a technology commonly used as a first step in manure and anaerobic digestate processing. After the S/L separation, a liquid fraction (LF) with soluble components, such as mineral N, potassium and other ions, is obtained. A solid fraction (SF) containing most of the organic matter and a significant proportion of P is also generated. Therefore, this technology allows nutrient and carbon distribution between the LF and SF, thus reducing management costs and enabling nutrient transport over longer distances (Kovačić et al., 2022; Jorgensen and Jensen, 2009). Separation mainly occurs via natural settling or mechanical methods (Bonmatí-Blasi et al., 2020; Prenafeta-Boldú and Parera, 2020). During natural settling, particles precipitate to the bottom of the manure pit, while the LF is extracted from the top (Lyons et al., 2021). This is a common process in the field of urban wastewater treatment, but it does not have many applications in manure management as it requires long retention times and large ponds, free of turbulence. A more effective separation can be obtained using mechanical methods which are able to achieve high removal efficiencies and a solid stream with high dry matter content and most of the nutrients (Prenafeta-Boldú and Parera, 2020).

The most common mechanical separation technologies can be divided into three main groups (Bonmatí-Blasi et al., 2020; Prenafeta-Boldú and Parera, 2020; Møller et al., 2000):

- Gravity (pore diameter 250-750 µm), which includes statics, vibratory and rotary sieves. In this case, solids cross the sieve via gravity, so no additional energy is required to achieve separation, which makes it the most energy-saving separation technology, and is typically used together with other separation systems.
- Pressure (pore diameter 300-650 µm), by screw pressing and filter belts; additional pressure is applied in order to increase the separation efficiency through the sieve.
- Centrifugation, using decanting centrifuges.

Over the last few decades, many studies have focused on comparing the efficiency and performance of different separation technologies (Prenafeta-Boldú and Parera, 2020; Møller et al., 2000; Flotats Ripoll et al., 2011). The percentage of dry matter and the nutrient content recovered in the SF are typically used as indicators of separation efficiency (Lyons et al., 2021; Møller et al., 2000). The most common methods for measuring efficiency are the simple separation efficiency index (SSEI) and the reduced separation efficiency index (RSEI). The SSEI is calculated by considering the nutrient and solids recovery, in terms of mass, as a proportion of the input material, so it indicates the proportion of a certain compound in the SF. An SSEI index of 0.5 would mean 50% of the component has been moved to the SF. On the contrary, the RSEI considers the difference in the masses of both fractions, showing the increase or decrease in concentration in the SF relative to the starting product (Lyons et al., 2021; Cocolo et al., 2012).

In order to determine the nutrient distribution between the two fractions, it is necessary to know the flow rate of the input manure and the output flow rate of both the SF and LF, as well as quantifying their nutrient content. Therefore, the factors affecting the efficiency of nutrient recovery in the SF must be considered in the decision-making process, and include the characteristics of the separation equipment, the size of the farm, the type of livestock, the investment and maintenance costs, ease of handling, adaptability to different types of manure, processing capacity, level of odour reduction (Møller et al., 2002) and the use of additives such as coagulant-flocculants or polyelectrolytes, which improve the separation efficiency, especially during centrifugation (Flotats Ripoll et al., 2011) (Table 1). These additives should not affect the quality and subsequent use of the separated SF as a fertilizer or organic amendment (Beggio et al., 2022; Prenafeta-Boldú and Parera, 2020).

Table 1 Reference values of separation efficiencies in terms of mass and nitrogen recovery in the solid fraction (%), depending on the type of livestock and separation technology

Separation technology		Fattening (>7% DM)		Sows (>3% DM)		Cattle (>5% DM)	
		Mass	Nitrogen	Mass	Nitrogen	Mass	Nitrogen
Gravity	Without additive	15	20	10	15	15	20
	With additive	20	25	15	20	20	25
Pressure	Without additive	15	20	10	15	20	25
	With additive	25	30	15	20	25	30
Centrifugation	Without additive	14	28	10	30	–	–
	With additive	20	50	15	45	20	50

Translated from Prenafeta-Boldú and Parera (2020).

Tables 1 and 2 show data from a review of manure treatment technologies carried out by the Government of Catalonia and the Institute of Agrifood Research and Technology (IRTA) in Spain (Prenafeta-Boldú and Parera, 2020). Table 1 compares the efficiency of the three main groups of separation techniques as a function of the type of livestock and the use of coagulant or flocculant; the highest efficiency is achieved using a centrifuge together with an additive. However, it should be noted that the associated costs in comparison to pressure technologies are significantly higher. Investment costs can range from €10 000 for sieves to €150 000 for a decanter centrifuge, while operational costs are mostly influenced by the use of additives (Flotats Ripoll et al., 2011). In addition, Table 2 shows a comparison of two of the most common separation technologies regarding nutrient (N, P) and metal content in the different fractions.

Fertilizer characteristics of the SF and LF, and hence the availability of nutrients for plants, are affected by the separation technique employed (Fangueiro et al., 2015) (Table 1). Furthermore, the inclusion of a previous process (e.g. anaerobic digestion or acidification) will also greatly influence nutrient speciation, removal efficiencies and the characteristics of the two streams obtained (Regueiro et al., 2016; Møller et al., 2000).

Prior to the solid-liquid separation, acidification minimizes NH_3 emissions during the solid-liquid separation and increases the P content of the liquid fraction (Cocolo et al., 2012; Regueiro et al., 2016). Both fractions can be used directly as a fertilizer, but it should be noted that the composition varies after

Table 2 Comparison of the nutrient content (nitrogen, phosphorus and potassium) and heavy metals (copper and zinc) in the different fractions, after a solid-liquid separation using a centrifuge with coagulants and polymers and a sieve and/or screw press separation system

Parameter	[a]Centrifuge + additive			[b]Sieve + screw press		
	Manure (kg/m³)	Liquid fraction (kg/m³)	Solid fraction (kg/t)	Manure (kg/m³)	Liquid fraction (kg/m³)	Solid fraction (kg/t)
Total nitrogen	3.37	1.93	9.24	4.30	4.10	6.07
Organic nitrogen	1.12	0.27	5.86	1.29	1.14	3.30
P	0.56	0.07	3.16	1.46	1.29	4.10
K	0.24	0.20	0.20	0.24	0.23	0.23
Cu (mg/kg)	4.88	0.66	30.43	15.16	14.70	23.27
Zn (mg/kg)	26.00	3.25	153.26	124.54	119.43	200.30

Translated from Prenafeta-Boldú and Parera (2020).
[a] Samples collected from one livestock farm with a centrifuge treating a mixture of pig and dairy manure ($n = 6$).
[b] Samples collected from seven livestock farms with a sieve and/or screw press separation system treating pig manure ($n = 17$).

separation and thus agronomic management should be modified accordingly (Table 2) (Prenafeta-Boldú and Parera, 2020).

Regarding the SF, the organic matter content (between 20% and 30%) provides a readily available carbon source (Camilleri-Rumbau et al., 2021) and, due to its high C/N ratio, it is more appropriate to be used as an organic amendment or in basal dressing; whereas the LF, with a lower C/N ratio (dry matter content of around 2–6%), is more suitable to be applied as a top dressing fertilizer, in fertigation or in soils with high phosphorus content, since the N/P ratio is higher than in the raw manure (Kovačić et al., 2022; Prenafeta-Boldú and Parera, 2020).

In parallel to management as fertilizer products, both fractions should be further processed in order to maximize nutrient recovery and/or to produce bioenergy. The solid fraction can be treated with consolidated technologies such as anaerobic digestion (see Section 2) and composting (see Section 4), or with other biomass valorization processes such as combustion, pelletization or pyrolysis. In addition, since solid-liquid separation does not guarantee high recovery of the nutrients still available in the obtained LF, a further post-treatment of the diluted LF is usually required (Camilleri-Rumbau et al., 2021). In this case, apart from the use of consolidated technologies such as anaerobic digestion or nitrification/denitrification (that cannot be considered a technology to recover but to remove nitrogen), other technologies such as stripping/absorption, struvite precipitation, membrane filtration or bioelectrochemical systems, in many cases combined with each other, can be implemented to maximize nutrient recovery from the LF, mainly in the form of struvite and ammonium sulphate (see Sections 5–8). These treatments result in different products that should be adequately managed taking into account their characteristics and chemical composition.

4 Composting

Composting can be defined as a nature-based solution for transforming solid organic wastes into compost, which is an organic amendment for agricultural soils, gardening and soil remediation (Cáceres et al., 2022); other specific compost products can be used as a growing medium for containerized plant production in certain proportions. The transformation of organic matter and other nutrients during composting is achieved by aerobic microorganisms under controlled conditions. Composting is usually carried out by piling up the solid organic matter (or a mixture with a bulking agent). The pile of organic material (composting pile) can be placed outside, with a cover structure, or composting can be achieved indoors within special tunnels (closed systems).

The primary field/lab parameters that should be measured to monitor the process are: temperature, oxygen concentration within the composting pile and moisture of the material. The temperature of the pile rises, as composting releases heat as a result of aerobic microbial activity (Haug, 1993; Cáceres et al., 2015a). Although oxygen availability is key in composting, water availability is necessary to maintain microbial activity, and optimize organic matter breakdown and compost maturity. Oxygen within the composting matrix decreases due to microbial consumption and the release of CO_2 as a product of respiration; this exhausted air should be renewed by fresh air. Therefore, the actions to manage the process are mainly aeration and irrigation of the material.

There are basically two types of composting systems that can be used in manure composting (i.e. static and dynamic composting) and they are distinguished by the way in which air is provided to the solid material. Air renewal (and also homogenization of the composting matrix) can be achieved by turning the material, and this is the key concept of dynamic composting. A static composting system consists of blowing air from the bottom of the composting piles through pipes connected to a blower (Cáceres et al., 2006). Static systems can be used outdoors, employing a simple configuration; but they are typically used in tunnel systems, which are closed structures where emitted gas can also be collected and treated to avoid the release of undesirable gases into the atmosphere.

Raw materials in livestock composting are mainly: manure from cattle or dairy cows, the solid fraction of pig or cattle slurry, droppings from pig or cattle manure (from organic livestock production), poultry manure and droppings from other animals under livestock production. These products are characterized by a high moisture content and high N and P content. Solid excrement is usually mixed with urine, which is why the manure has a high water content. A high moisture content (50-70%, fresh weight basis) is necessary to trigger the composting process, which is facilitated by a quick temperature rise. Other products, like the solid fraction of cattle or pig slurry, could have a very high moisture content (depending on the physical structure of the material) and should be mixed with bulking agents to diminish the water content of the mixture. When the solid fraction is obtained from slurry, the structure of the material and the moisture content sometimes hinders piling up the material making it necessary to separate the slurry combining two systems (such as gravity and a screw press).

Solid manure is characterized by the presence of bedding materials used for animal welfare (straw or wood-based materials such as sawdust). Bulking agents are normally added to other raw materials during composting; however, for livestock manure, the addition of such materials is not usually necessary, although in some cases it would be convenient in order to increase the C/N ratio and avoid certain ammonia and greenhouse gas production. The emission of such gases are issues of concern, but mitigating this release can be achieved by raising the C/N ratio of the manure or the mixtures by adding bulking agents. At lab level, it has been demonstrated that the addition of biochar can diminish the gases released into the atmosphere (Czekala et al., 2016; Janczak et al., 2017).

The solid fraction of cattle or pig/cattle slurry normally has a lower N content than the raw slurry, which is due to part of the N in the slurry remaining in the liquid portion, after separation. In general, the solid fraction of cattle slurry is less concentrated than any other kind of manure (Table 3).

The chemical and physical characteristics of the solid fraction of cattle or pig slurry can be significantly improved by adding a bulking agent: the C/N increases, the moisture content diminishes, the organic matter increases and the N decreases. On the other hand, the physico-chemical parameters (pH and EC) diminish. Table 3 shows the effect of the bulking agent (pine debris) addition, improving the air capacity of the mixture. The study by Cáceres et al., tested the composting of just separated cattle slurry using dynamic composting, separated cattle slurry using a static method for composting and separated cattle slurry with pine debris also composted dynamically. Results show that the evolution of several chemical parameters and physico-chemical parameters differed, in particular, regarding electrical conductivity, ammonia-N and nitrate-N (the different evolution of this N species had a clear consequence on pH as nitrification naturally drops the pH) (Fig. 4, Cáceres et al., 2006).

Over the last few years, composting has been studied as a process of processes (Cáceres et al., 2018). In particular, the fate of N during composting is of interest due to the enrichment of N in the compost, the loss of this element in the form of NH_3 at the beginning of the process and the loss of N_2O due to other processes during composting (Cáceres et al., 2018). Specifically, it has been demonstrated that nitrification, which is a process that can occur in soil and other liquid treatment processes, is also possible during composting under certain conditions (Cáceres et al., 2006, 2016).

Table 3 Characterization of the solid fraction of cattle slurry (SF-CS) without and with pile debris (PD) at 2/1, v/v rate (Cáceres et al., 2006)

Parameter	Units	SF-CS	SF-CS/PD (2/1, v/v)
Gravimetric moisture	%, fb	82.0	72.1
Organic matter	%, db	92.4	95.5
Norg	%, db	1.4	1.1
C/Norg	–	39	50
pH	–	7.7	7.2
Electrical conductivity	dS/m	0.80	0.7
Total porosity	% vol.	95.9	94.8
Volumetric moisture	% vol.	31.9	21.9
Air capacity	% vol.	64.0	72.9

Adapted from: Cáceres et al. (2006).
db, dry basis; Fb, fresh basis; Norg, organic N.

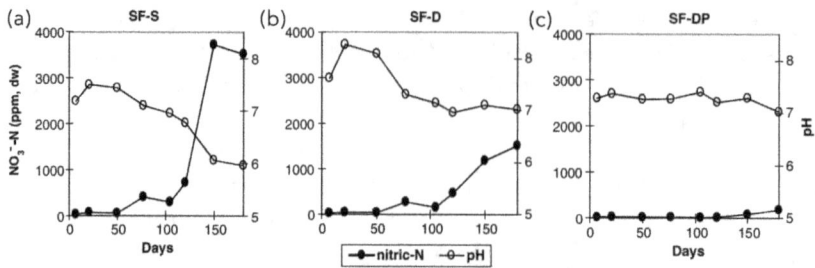

Figure 4 Time course of nitrate-N and pH during the composting of the solid fraction of cattle slurry: (a) solid fraction using a static method (SF-S), (b) solid fraction using a dynamic method (SF-D) and (c) solid fraction with pine debris (2/1, v/v) using a dynamic method. Source: Figures extracted from Cáceres et al. (2006).

Composts based on manure are, in general, rich in plant nutrients but also in other undesirable elements for plant growth, such as sodium which increases the electrical conductivity of the product (Cáceres et al., 2015b). Composts of the solid fraction of pig or cattle slurry are suitable for many applications and the fate of N (i.e. organic N, ammonium N and nitrate N) should be studied for each climate condition (Cáceres et al., 2022), and in vulnerable areas, it is a simple and advisable way to manage slurries. Separators (see Section 3) can be purchased or rented, and the liquid can be applied in cultures near the farms, while the solid fraction can be transformed into compost on the farm (*on farm composting*) and in this way the compost (reduced by 50% in volume or weight with respect to the initial mixture) can be used in other locations. Composting can also be used in combination with anaerobic digestion. The solid part of the digestate can be composted in order, for example, to transform N-ammonia, which is the prominent mineral form in such a by-product, into N-nitrate. Furthermore, organic matter can be transformed even further under aerobic conditions before applying the product (digestate) to the soil.

Varying actions can be applied to promote the composting of different animal manures by using, for instance, alternative bulking agents. The Catalan Government recently published a manual for on-farm composting of manure, which includes the characterization of different types of manures and bulking agents, including aspects such as using the generated leachates or the commercialization of compost (Cáceres and Parera, 2022). Indeed, important amounts of leachates can be generated depending on the management of the composting piles and the climatic conditions. Hence, new approaches to the transformation (via nitrification) of these effluents and their use in horticultural production have been proposed (Cáceres et al., 2015c).

5 Struvite precipitation

Struvite precipitation has been described as the most promising and universal solution for phosphorus recovery from livestock manure and anaerobic digestates. Struvite, or magnesium ammonium phosphate (MAP, $MgNH_4PO_4 \cdot 6H_2O$), is a white crystalline element (Fig. 5) composed of ammonia nitrogen (NH_4^+), phosphorus (phosphate, PO_4^{3}) and magnesium (Mg^{2+}) in equimolar relation. Struvite precipitation is achieved under alkaline conditions when the concentration of Mg^{2+}, NH_4^+ and PO_4^{3-} exceeds the solubility product, according to the following reaction (Cerrillo et al., 2015):

$$Mg^{2+} + NH_4^+ + H_2PO_4^- + 6H_2O \rightarrow MgNH_4PO_4 \cdot 6H_2O + 2H^+$$

Struvite precipitation is dependent on many factors, a principal parameter being pH, as it has a direct effect on the concentration of free ions available for reaction and on struvite solubility. A wide range of optimal pH values have been reported for struvite precipitation in the literature; ranging from 7 to 11 with an optimal at 8.5-9.3 depending, for instance, on the feedstock characteristics and operational conditions (Nagarajan et al., 2023). Moreover, since struvite production also generates H^+, the continuous addition of alkali (generally using NaOH or $Mg(OH)_2$) is needed to keep the pH in the optimal range and to favour struvite precipitation. Besides pH, the N/P and Mg/P molar ratios, the physical characteristics of the reactor (stirring and settling properties), the temperature and the ionic strength of competitive cations such as Ca^{2+} and Na^+, are other factors affecting struvite precipitation (Tao et al., 2016; Cerrillo et al., 2015).

On the other hand, high NH_4^+ concentrations are usually found in anaerobic digestates in comparison with PO_4^{3-}, but as Mg^{2+} is insufficient, supplements such as $Mg(OH)_2$, $MgCl_2$ or $MgSO_4$ have to be added to the reactor in order

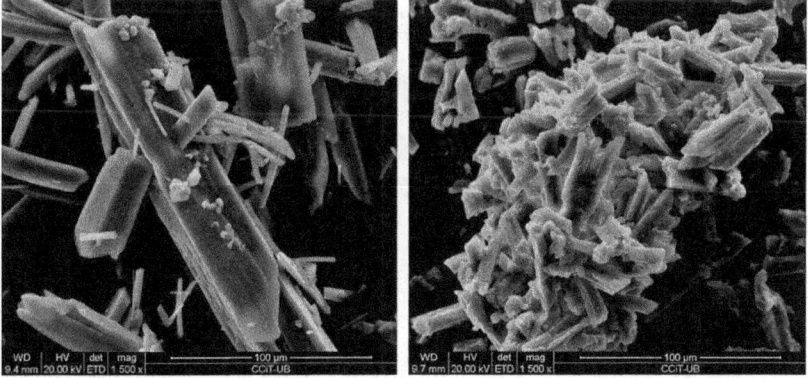

Figure 5 Struvite crystals obtained from pig slurry (left) and synthetic wastewater (right). (Bonmatí ©).

to induce struvite formation and achieve optimal molar ratios (Nagarajan et al., 2023; Lorick et al., 2020; Cerrillo et al., 2015). It should be noted that the co-precipitation caused by Ca^{2+} (as calcium phosphate), which is usually highly concentrated in raw dairy manure, would negatively affect the recovery yield of struvite. In this sense, the highest recovery rates have been reported from studies on the treatment of the liquid fraction of manure-derived digestates after a mechanical solid-liquid separation (Brown et al., 2018). However, in many cases, the major obstacle for P recovery as struvite is due to 55–65% of the total amount of P from manure digestate being concentrated in the solid fraction (see Section 3), so P should be previously recovered in the liquid fraction as soluble inorganic phosphorus. Therefore, it may be necessary to carry out pre-treatment to maximize phosphorus recovery. According to the literature, acid treatment would be the best pre-treatment for the release of phosphorus (Shi et al., 2018).

In a meta-analyses of different livestock treatment technologies, struvite was shown to be the most promising for P recovery, achieving the highest recovery among all the evaluated technologies with a rate of 83.0% (Shi et al., 2022). In this sense, struvite precipitation has been extensively studied and applied to anaerobically digested dairy and pig manure (Wagner and Karthikeyan, 2022; Cerrillo et al., 2021; Tao et al., 2016; Cerrillo et al., 2015) poultry manure (Yetilmezsoy and Sapci-Zengin, 2009), or the liquid fraction of pig slurry (Huang et al., 2011), reporting PO_4^{3-} recovery efficiencies of up to 95%. The typical struvite precipitation reactor consists of a stirred crystallization reactor composed of a reaction zone and a sedimentation zone (Fig. 6). The reactor could also include an aeration system to favour CO_2 stripping and pH increase, and prevent mechanical mixing (Cerrillo et al., 2015). However, other innovative approaches for P recovery as struvite, such as bioelectrochemical and membrane systems, with different configurations, have been gaining attention over the last few years (Cerrillo et al., 2023; Cerrillo et al., 2021).

In relation to its fertilizer quality, struvite is considered to be an effective and clean slow-release fertilizer, as it is sparingly soluble under neutral and alkaline conditions but readily soluble in acid (Carreras-Sempere et al., 2021; Cerrillo et al., 2021). Nagarajan et al. (2023) reported that 1 kg of MAP per day is enough to fertilize 2.6 hectares of arable land. The value of struvite as fertilizer has only recently been understood and it is now the focus of increasing research attention (Robles-Aguilar et al., 2019, 2020). The advantages of struvite as a solid fertilizer include ease of application, low transportation costs, reduced bulk storage, and the avoidance of odour and pathogen contamination issues (Tao et al., 2016). In addition, P recovery in the form of struvite reduces the use of P rock-based mineral fertilizers, allowing the release of nutrients at an appropriate rate for crop uptake due to its low solubility, while reducing nutrient run-off and leaching (Nkoa, 2014). Struvite has also been applied to nutrient

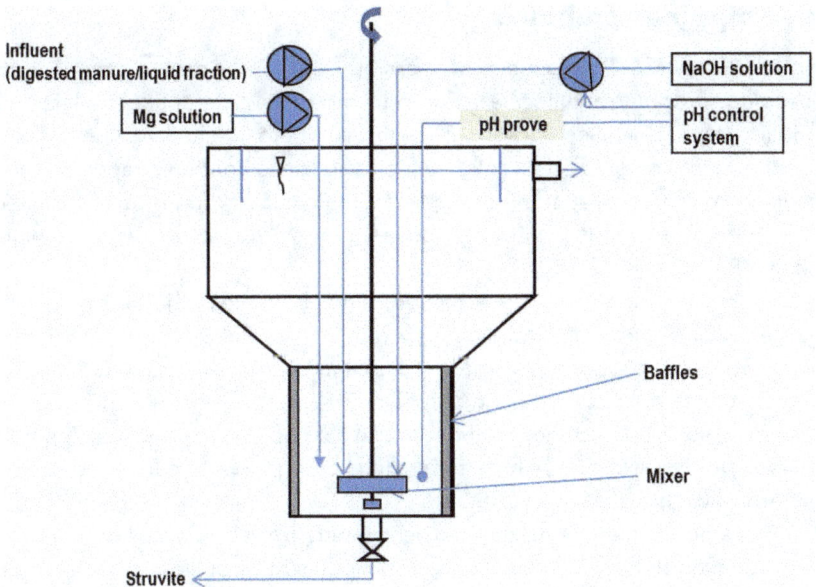

Figure 6 Schematic diagram of a crystallization lab-scale reactor. Source: Cerrillo et al. (2015).

solutions in horticulture (without an additional P source), attaining similar plant performance (Carreras-Sempere et al., 2021, 2022).

Struvite precipitation has been implemented in more than 40 industrial plants around the world (Rizzioli et al., 2023). However, this process is typically applied for wastewater and municipal sludge digestate (Pikaar et al., 2022; Sena and Hicks, 2018), with some commercial full-scale systems already on the market (Ostara Pearl®, NuReSys®, PHOSPAQ™, Airprex®) (Rizzioli et al., 2023; Bonmatí-Blasi et al., 2020). However, there is limited information for agricultural digestate, due to the low P content, which has a direct bearing on process efficiency. In addition, P recovery from agricultural residues is not often applied as there is not a specific regulation focused on this nutrient, as is the case for N by means of the Nitrate Directive. However, since agricultural residues are considered to be a critical raw material, some countries such as The Netherlands, Denmark, Belgium and Germany have already established special authorization for the use of struvite recovered from manure as a fertilizer (Corona et al., 2021). An example of a prototype demonstration in an operational environment (TRL7) is the REVAWASTE® system (CARTIF company, Spain), at demonstrative scale, that treats agricultural residues using the AD process, including a crystallization unit that allows nutrient recovery of up to 95.4% (Rizzioli et al., 2023).

6 Stripping/scrubbing

Sigurnjak et al. (2019), described the possible pathways to recover ammonia via stripping-scrubbing at full scale in Europe. Among the different possibilities, we will discuss the so-called end-of-pipe pathway, where livestock manure is firstly separated into a liquid and a solid fraction and, since ammonia represents the major nitrogen compound of the liquid fraction, this fraction is further treated in a stripping-scrubbing unit to produce ammonium sulphate or ammonium nitrate (Rizzioli et al., 2023).

Ammonia stripping can be achieved using air, steam or biogas to separate the gaseous NH_3 from the liquid phase (Adghim et al., 2023). The NH_3-saturated stripping gas subsequently comes into contact with an acid solution, usually against the flow, to capture ammonia (Fig. 7). When sulphuric acid is used, a $(NH_4)_2SO_4$ solution is formed. Nitric acid can also be used to capture ammonia that will produce NH_4NO_3, which is more interesting for fertilization, but is more expensive (Bonmatí et al., 2020).

The amount of ammonia that can be stripped from liquid waste or absorbed in the acidic solution is based on two thermodynamic equilibria, which are pH and temperature dependent (Bonmatí-Blasi et al., 2020):

- Ammonia dissociation equilibrium in the liquid: $NH_{3\,(l)} \leftrightarrow NH_3$ (g); and
- Ammonia gas/liquid equilibrium: $NH_4^+ \leftrightarrow NH_3 + H^+$.

For instance, at room temperature (about 20°C), ammonia stripping requires a high pH to move the ammonia equilibrium towards free ammonia. However, if the temperature increases, a lower pH is needed. As previously mentioned, pH

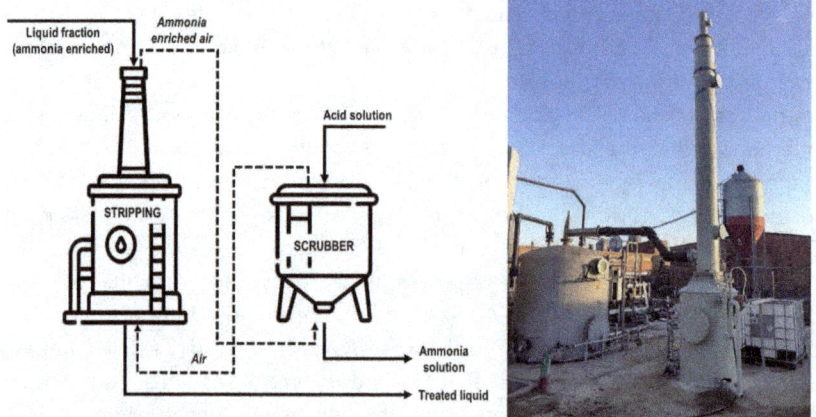

Figure 7 (left) Schematic diagram of the stripping/scrubbing process and (right) image of equipment installed at an existing biogas plant in Catalonia, Spain (own picture).

is usually set between 9 and 10 by means of base addition (lime is frequently used due to its low cost) or previous CO_2 stripping. Nevertheless, if air stripping is performed at high temperature, the high manure buffering capacity could enable maintaining the pH at the required value, and the amount of alkali could be reduced. On the other hand, pH values below 5 highly favour ammonia scrubbing, with almost no temperature effect (Laureni et al., 2013; Bonmatí and Flotats, 2003). Thus, one of the main limiting factors for ammonia air stripping, which is usually performed at high temperature, is the availability of an economical thermal energy source. In this regard, by combining anaerobic digestion and a stripping/absorption process, the biogas generated could provide the input thermal energy needed for stripping at high temperature (Shi et al., 2018; Bonmatí and Flotats, 2003).

In terms of efficiency, up to 95% of ammonia reduction can be achieved by setting optimal conditions and almost complete ammonia recovery is possible via absorption in acid solutions with only a little acid stoichiometric excess (Bonmatí-Blasi et al., 2020). Shi et al. (2018) compared several studies on NH_3 stripping from pig manure, dairy manure, chicken manure and food waste. Among the different conditions and feedstocks, the higher efficiencies (greater than 90%) were achieved when treating pig manure and keeping the pH above 9.5. Moreover, when combined with prior anaerobic digestion, the performance of the stripping process can be altered, as the biological process increases the ammonia content, part of the organic nitrogen is hydrolyzed to ammonia, which slightly increases the pH and reduces alkali requirement (Bonmatí-Blasi et al., 2020). Bonmatí and Flotats (2003) studied the differences between the stripping performance when using digested or fresh pig manure. For fresh slurry, despite working at 80°C, a high initial pH (11.5) was required to achieve ammonia removal at approximately 63%. On the other hand, for digested pig manure, 90% of ammonia was removed without pH modification and there was significantly less organic matter contaminating the recovered ammonia salt. It was also shown that substrates characterized by low organic matter content resulted in ammonia stripping efficiencies greater than 80% (Laureni et al., 2013).

The ammonium salt produced after the scrubber can be used as a biobased fertilizer, thus representing a valid alternative to fossil-based fertilizers, as it contains total N entirely in mineral form, similar to synthetic mineral N fertilizers, produced via the Haber–Bosch process. However, the high variability of N concentrations in these products is still the biggest challenge for their recognition as N fertilizers (Sigurnjak et al., 2019). As mentioned previously, the current European legislative framework prohibits application of these end-products above 170 kg N ha^{-1} y^{-1}, which reduces the market opportunities for the installation owners and hinders the development of a circular economy. Fortunately, the new Fertilizing Product Regulation 1009/2019 and the technical proposal of manure-derived RENURE criteria could open new alternatives for the use of NPK biobased fertilizers (Brienza et al., 2021).

Brienza et al. (2021), analysed the performance of vacuum NH_3 stripping and a scrubbing system that allowed the recovery of 57% of ammonia present in the digestate, in the form of an ammonium sulphate solution. The characteristics of the recovered product (22% solution) would fit the proposed RENURE criteria thus enabling the reuse of N derived from manure as inorganic biobased N fertilizer in the European market to replace mineral N fertilizer (Sigurnjak et al., 2019). However, future studies should further investigate the agronomical and environmental performance of ammonium nitrate and ammonium sulphate under varying conditions and over multi-year crop rotations.

Some examples of commercially available stripping technologies are the AMFER® stripping system (Colsen industry, The Netherlands), the Biogas Bree system (Bree, Belgium), the 'Detricon' (Gistel, Belgium) and the 'NITROStripp' system (Chiari, Italy). The operational costs associated with this process range between €2 and 7/kg N removed, primarily depending on the reagent used for pH regulation, the temperature and energy consumption (Rizzioli et al., 2023).

7 Membrane filtration

Membrane filtration is a physical process in which a membrane acts as a filter thus separating a liquid stream into a solid fraction, also known as concentrate or retentate, and into a liquid fraction, called a permeate, that are not thermally, chemically or biologically modified. Membrane filtration is usually applied to the manure or digestate liquid fraction after mechanical separation (Rizzioli et al., 2023). However, depending on the separation system, a second separation process is sometimes needed in order to avoid damaging the membrane modules due to the high solid content of the liquid fraction (see Section 3). Membrane filtration has gained interest in recent years in the agri-food sector, especially in surplus nutrient areas with high livestock density, as it concentrates manure nutrients in small volumes that could be exported as fertilizers to other agricultural regions.

Membranes are typically classified depending on their pore size, applied pressure, molecular weight cut-off and permeability. These mainly include hydraulic pressure membrane processes such as microfiltration (MF; pore size: 0.1-5 μm), ultrafiltration (UF; pore size: 0.005-0.1 μm), nanofiltration (NF; pore size: 0.001-0.01 μm), reverse osmosis (RO; pore size: 0.0001-0.001 μm), and other emerging membrane technologies such as forward osmosis (FO), membrane distillation (MD) and electrodialysis (ED) (Pinnekamp and Friedrich, 2006).

Masse et al. (2007) performed a critical review of the current state of research on the use of membrane filtration for manure concentration and treatment. MF and UF membranes basically act as very efficient solid-liquid separators that can isolate nutrients associated with particles such as phosphorus. Furthermore, MF or UF membranes can successfully eliminate particles from digestate to achieve over 80% removal of organic matter, when combined with paper filtration, and UF up to 95.7% of suspended solids can be removed from digested swine manure (Zhang et al., 2020).

Ammonia and potassium retention requires NF or RO and represents one of the major challenges of using RO filtration for animal manure treatment. Konieczny et al. (2011) recovered water from pig slurry by implementing an integrated system consisting of centrifugation followed by UF and NF. Guo and Jin (2014) investigated anaerobically digested cattle manure treatments with RO pre-treated with UF, the concentration of NH_4^+ in the final permeate was around 63 mg/L from the initial 836 mg/L. Rejection of other species such as K^+, Na^+ and Cl^- was observed to be as high as 90%.

Table 4 shows a summary of these membrane processes and operational characteristics.

However, depending on the type of membrane used, there are several technical challenges for the application of membrane processes at an industrial level, such as membrane fouling, contaminant enrichment, ammonia volatilization and high maintenance costs, (Zhang et al., 2020). Fouling seems to be one of the main problems affecting NF and RO since it impairs membrane performance and shortens membrane lifetime, thereby restraining the productivity of nutrient recovery, due to the high concentrations of various foulants, such as humic acids, suspended solids, colloids and inorganic salts, which are nutrients in digested effluent (Zhang et al., 2020; Velthof, 2011). Thus, the maximum amount of solids and organic matter should be removed from the liquid fraction prior the application of RO by applying pre-treatments that include techniques such as UF, dissolved air flotation, low pressure membrane filtration and manure/digestate coagulation-flocculation, among others (Zhang et al., 2020; Masse et al., 2007).

Figure 8 shows the flowchart of a pig manure processing plant located in The Netherlands, which includes anaerobic co-digestion, solid-liquid separation and further treatment of the liquid fraction by ultrafiltration, followed by reverse osmosis trying to overcome RO membrane fouling.

However, these pre-treatment techniques can reduce the nutrient content in the digestate, and subsequent nutrient concentration, and require additional investment for residual treatment. In this sense, forward osmosis, membrane distillation and electrodialysis are three emerging membrane-based processes considered suitable to overcome the challenges of nutrient recovery, which could potentially represent a paradigm shift in nutrient management (Zhang et al., 2020). In FO, the driving force of the process is the chemical potential difference across the membrane, instead of hydraulic pressure, which forces

Table 4 Membrane process classification and operational conditions

Membrane process	Pore size (μm)	Size of the particles, colloids or molecules to be separated	Driving force (pressure difference, bar)	Application
Microfiltration	0.1–5	Solids > 0.1 μm	0.1–3	Separation of solid matter from suspensions
Ultrafiltration	0.005–0.1	200000–20000 D[a]	0.5–10	Separation of macromolecules or colloids, disinfection
Nanofiltration	0.001–0.01	20000–200 D[a]	2–40	Separation of dissolved organic molecules and polyvalent inorganic ions
Reverse osmosis	0.0001–0.001	<200 D[*a]	5–70 Up to 120 in special cases	Separation of organic molecules and all ions

[a]Dalton, numerically equivalent to the molecular weight in g/mol (modified from Pinnekamp and Friedrich, 2006)

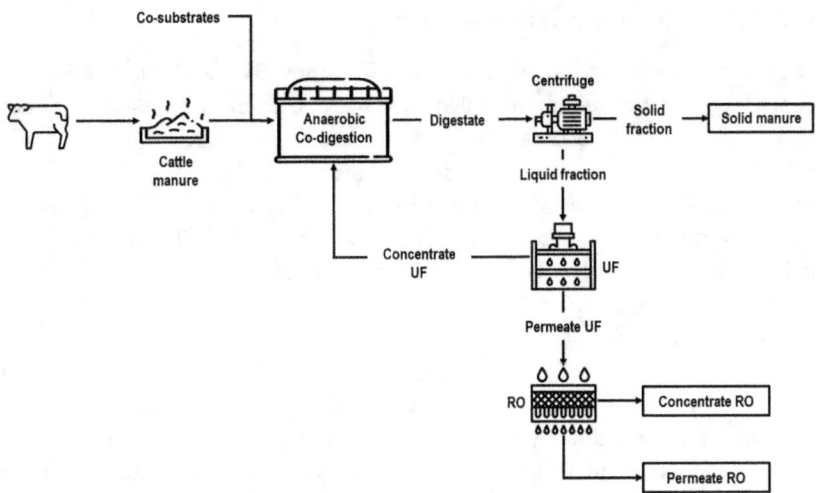

Figure 8 Treatment scheme of a manure treatment plant located in The Netherlands. Source: Modified from Velthof, 2011. https://edepot.wur.nl/192069.

the liquid to flow from the low to high concentrated compartment. FO has been demonstrated to be less sensible and more reversible to fouling than RO (Blandin et al., 2016).

Using hydrophobic membranes for MD, the driving force is based on the volatility and vapour pressure of the different compounds that are to be separated. Since the process is driven by temperature gradients, residual heat can be used, and the use of expensive pressure pumps is unnecessary. Recently, hydrophobic membranes permeable to gases have been applied to ammonia recovery from manure with recovery efficiencies of more than 97%. Furthermore, hydrophobic membranes can be combined with other treatment technologies such as anaerobic digestion (see Section 2) to increase biogas production, or with bioelectrochemical systems to recover ammonia (see Section 8) (Cerrillo et al., 2023; Cerrillo et al., 2021).

Finally, ED is an electro-membrane process in which ions are transported through an ion exchange membrane under the influence of a potential gradient. In ED, the direct current field forces the migration of cations and anions toward cathodes and anodes, respectively (Zhang et al., 2020). The ion separation is achieved via ion-exchange membranes that include cation-selective, anion-selective and bipolar membranes that serve as a barrier to fraction these ions (Zhang et al., 2020). Therefore, the excess nitrogen and phosphorus, which are found as ions in nutrient-rich streams, such as swine liquid fraction, can be removed using this technique (Fukumoto and Haga, 2004).

8 Bioelectrochemical systems

Bioelectrochemical systems (BESs) are an innovative and highly versatile technology for the treatment of livestock manure to obtain biofertilizers and recover bioenergy. These devices are bioreactors that include an anode and a cathode, and that use microorganisms to catalyse oxidation and/or reduction reactions. Several organic substrates have been used in BESs, e.g. livestock manure or digested livestock manure (Cerrillo et al., 2016). The organic matter of the substrate is oxidized by exoelectrogenic bacteria that grow on a biofilm on the bioanode, and the produced electrons circulate through an external electrical circuit to the cathode.

BESs have different applications in the field of livestock manure treatment. First, electron movement through the external circuit produces electricity, which can be used directly to power small devices via the chemical energy contained in the organic substrate as an energy source, and has even been successfully boosted up to 99 V (Koffi and Okabe, 2020). In this case, the BES is called a microbial fuel cell (MFC).

Second, the movement of electrons can also be used to force the migration of ions through a membrane in a dual-chambered MFC. In this case, an anode compartment and a cathode compartment are divided by a cation or an anion exchange membrane. This way, the electron circulation from the anode to the cathode generates a charge imbalance in the cell, which is compensated by ion migration through the membrane, from one compartment to the other. Using a cationic exchange membrane in a dual-chambered BES, ammonium can be recovered from livestock manure, which is introduced into the anode compartment, in a clean solution placed in the cathode compartment (Sotres et al., 2015). The livestock manure treated in the anode compartment will have a lower organic and nitrogen content, while the cathodic solution, rich in ammonium, can be used as a fertilizer. Since the migration of ammonia is related to the amount of charge that is circulating between the anode and the cathode, ammonium recovery can be boosted with the aid of a small energy input, achieving a flux through the membrane of up to 119 gN m^{-2} day^{-1} (Rodríguez Arredondo et al., 2019). In a recent review, it has been reported that the energy consumption for ammonia recovery in BESs is in the range of 1.17–2.7 kWh kg$_N^{-1}$, generally only considering the external energy supplied to enhance ammonium migration (Cerrillo et al., 2023).

BESs have evolved for the simultaneous recovery of ammonium and phosphate, in the form of a concentrated solution or via salt precipitation as struvite, using a combination of different types of membranes or number of chambers (Fig. 9) (Haddadi et al., 2014; Cerrillo et al., 2021b, 2023; Cusick et al., 2014). For the recovery of phosphorus from livestock manure, a first step may involve solubilization from ferrous phosphate mineral, in order to allow

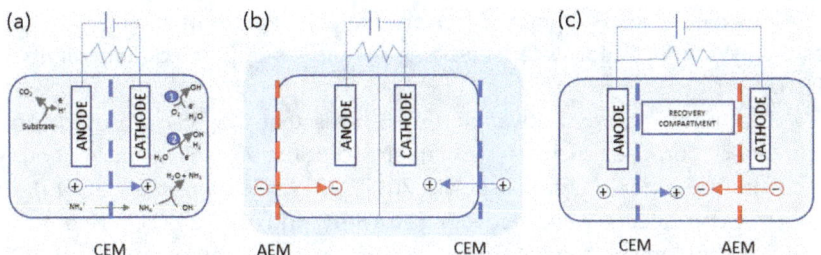

Figure 9 Three examples of typical configurations of BESs for nutrient recovery. + and − stand for ions charged positively (i.e. NH_4^+) and negatively (i.e. PO_4^{3-}), respectively. (a) Dual-chambered BESs, (b) submersed BESs and (c) three-chambered cell with recovery compartment. The main reactions taking place in BESs are shown in (a) by way of example, where ❶ and ❷ stand for MFC and MEC cathodic reactions, respectively (Cerrillo et al., 2023).

migration in the form of phosphate through the anionic membranes (Cerrillo et al., 2021b).

Third, by applying a small external voltage to BESs, certain non-spontaneous reactions can be favoured, which is the case for the production of biohydrogen or biomethane in microbial electrolysis cells (MECs). Hydrogen is produced via the combination of electrons that travel to the cathode with protons (eq. 1) (Cheng and Logan, 2011), while for the production of methane, carbon dioxide has to be introduced into the biocathode compartment, where bioelectromethanogenesis reactions occur (Zeppilli et al., 2020). Methane obtained from CO_2 in BESs can be achieved via two different mechanisms (Villano et al., 2011): (i) indirectly, through the microbially catalysed and/or intermediate abiotic electrochemical production of hydrogen in the cathodic compartment (Equations 1 and 2) or (ii) directly, via reducing CO_2 to methane by taking the necessary electrons from the cathode (eq. 3).

$$2H^+ + 2e^- \rightarrow H_2 \text{ (1)}$$
$$4H_2 + CO_2 \rightarrow CH_4 + 2H_2O \text{ (2)}$$
$$CO_2 + 8H^+ + 8e^- \rightarrow CH_4 + 2H_2O \text{ (3)}$$

Bioelectromethanogenesis application in BESs is of great interest for bioenergy production, as biomethane can be produced from carbon dioxide captured from different sources while treating livestock manure. One possible source of carbon dioxide is biogas obtained from anaerobic digestion, so MECs can be applied to biogas upgrading, increasing biomethane production thanks to the conversion of carbon dioxide, which differs from conventional biogas-upgrading technologies. A tubular MEC of 12 L, continuously fed by a gas mixture composed of CO_2 at 30% and N_2 at 70% to simulate the CO_2 content of a biogas, has been reported to achieve a methane production of 1.25 L day^{-1}. This study reported a low energy consumption of 0.33 kWh m^{-3} CO_2, resulting in a lower value than commercially available technologies on the market, such as 0.8 kWh m³ CO_2 for water scrubbing for biogas upgrading (Zeppilli et al., 2020).

The most important advantage of BESs is that they can be used for a combined purpose, such as biomethane production while simultaneously recovering ammonia (Zeppilli et al., 2017, 2019; Cerrillo et al., 2021a), or hydrogen production while recovering struvite (Song et al., 2021). Furthermore, it is a technology with great possibilities when combined with AD, either to recover nutrients or to increase biogas production. BESs can be submersed in anaerobic digesters (Zhang and Angelidaki, 2015), become a post-treatment for digestate (Cerrillo et al., 2016a), or be placed in a recirculation loop of the anaerobic digester to increase the stability of the system and biogas production (Cerrillo et al., 2016b). A recovery of 40% of the nitrogen in pig slurry in a combined AD-MEC system and an increase of 55% in methane production when the MEC was integrated in a recirculation loop in the AD has been reported (Cerrillo et al., 2016b).

Despite the great versatility and potential of BESs for biofertilizers and bioenergy recovery, most of the experience to date has been at lab or small pilot scale. More research is needed to improve nutrient and energy recovery efficiencies of BESs, with a low construction cost, in order to boost scaling up and commercialization of this technology.

9 Case study: farm for the future

Figure 10 shows an example of a circular approach based on a mixed farming system (ruminant production + fodder crop production) to be implemented in cattle farms, which includes not only some of the technologies described in this chapter, such as anaerobic digestion, solid/liquid separation or composting, but also strategies to increase biogas production and agronomical practices that allow reducing the environmental impact derived from the use of manure-based fertilizers in croplands.

The case study was built up in the framework of a LIFE project called 'Farms for the future: innovation for sustainable manure management from farm to soil (LIFE12 ENV/ES/000647)'. The main objective of this study was to investigate the nutrient extraction efficiency from soil of three different catch crop (CC) species, as well as the viability of using them as co-substrates for anaerobic co-digestion with dairy manure, with emphasis on the effect of ensiling on the anaerobic biodegradability and biogas potential. Digestate from the anaerobic co-digestion was used to fertilize the main crop, thus closing nutrient cycles. The study was published in the Waste Management Journal, and is available in open access (Riau et al., 2021).

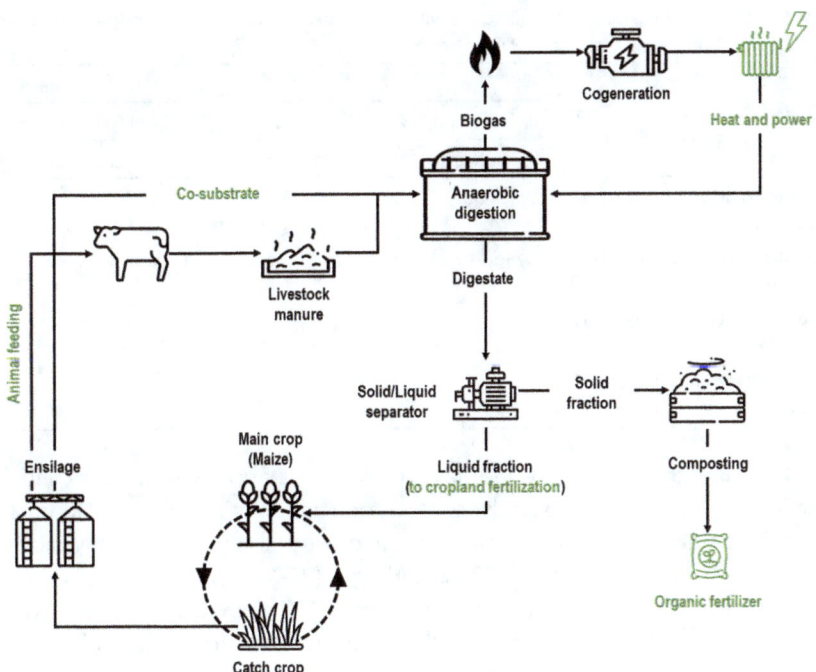

Figure 10 Circular approach in a cattle farm. Source: Riau et al. (2021).

CCs can be grown in the period between two main crops that have been fertilized with manure, anaerobic digestates and/or its liquid fraction, as a strategy to retain soil mineral N as well as P and heavy metals, thus reducing leaching and run-off losses and improving soil quality, by adding organic matter and reducing fertilizer requirements of the following growing season.

In this study, three CCs species were tested: *Lolium multiflorum* (ryegrass), *Brassica napus* (forage rape) and *Avena strigosa* (black oat). The CCs were sown once the main crop was harvested, by the end of September or the beginning of October, and were grown with minimum cultivation and harvesting effort as well as without using fertilizers and irrigation water. Finally, CCs were harvested in March. This rotation was repeated during three consecutive years. As an example, the experimental design of one rotation is summarized in Table 5. As a control of the process, a plot without CCs was also tested.

Despite spontaneous herbage growth, N uptake from soil via catch crops was always significantly higher (42–264%) than those cases in which CC was not sown, although highly variable among years. Thus, CC may reduce the risk of nitrate leaching from soil. Over different years, the highest N and P extractions were achieved by forage rape. On the other hand, it is known that anaerobic digestion does not affect the removal of dissolved Cu and Zn in dairy manure. However, Cu and Zn bioavailability is increased during the process, thus

Table 5 Experimental design of field trials with maize and catch crop rotations (plots: 131 m²; three repetitions each; three consecutive seasons)

Maize fertilization (N dosage = 170 kg N ha⁻¹)	Year 1												Year 2	
	Ja	F	M	A	M	Jn	J	A	S	O	N	D	Ja	F
	Catch crop			Main crop					Catch crop					
Liquid fraction of digested manure	Black oat		Maize							Black oat				
Liquid fraction of digested manure	Ryegrass		Maize							Ryegrass				
Liquid fraction of digested manure	Forage rape		Maize							Forage rape				
Liquid fraction of digested manure	-		Maize							-				

making anaerobic digestate less favourable for its application in agriculture. Nevertheless, this could be solved by introducing CCs into crop rotation. Zn and Cu uptake from all catch crops studied were always significantly higher than those obtained by spontaneous herbages (27-197%).

As mentioned in Section 1, once harvested, CC biomass can be processed as a co-substrate in the anaerobic digestion process, together with livestock manure, in order to increase biogas production since, on one side, manure provides buffering capacity and a wide range of nutrients, while on the other, CCs, with high carbon content, balance the carbon to nitrogen (C/N) ratio of the feedstock, thereby decreasing the risk of ammonia inhibition. However, since CCs are usually seasonally sown (autumn) and harvested (spring), it is necessary to find a storage method that guarantees the availability of this feedstock for biogas production during the whole year. This is why CCs were ensiled after harvesting, as it has been demonstrated that ensiling can maintain the energy content of crops (over 90%), ensuring good nutritional value when used as animal feed. Moreover, during ensiling, the different biochemical processes, such as hydrolysis and acidification, directly or indirectly affect biogas production by altering the feedstock properties. A faster pH decrease produces more water-soluble carbohydrates in the silage mass resulting in more biogas. Therefore, ensiling not only allows maintaining crop characteristics but, as a feedstock for biogas production, increases the anaerobic biodegradability of lignocellulosic materials thus improving process efficiency. In fact, the anaerobic biodegradability of the ensiled ryegrass, forage rape and black oat was increased by 10%, 14% and 10%, respectively, which resulted in an increase of the methane yield by 40%, 46% and 50%, in terms of LCH$_4$ per tonne of waste and 19%, 25% and 27%, in terms of methane production per volatile solid added.

Table 6 Organic loading rate, operational conditions and methane yields obtained at steady-state conditions for each semi-continuous experiment

	Units	Ryegrass		Forage rape		Black oat	
		Control	Co-dig	Control	Co-dig	Control	Co-dig
OLR	kgCOD m^{-3}·day^{-1}	1.8 ± 0.1	2.4 ± 0.4	1.2 ± 0.2	1.4 ± 0.4	1.3 ± 0.2	1.9 ± 0.4
VS removal	%	40 ± 19	56 ± 7	42 ± 15	57 ± 10	52 ± 10	57 ± 6
Methane content	%	61 ± 15	63 ± 0	59 ± 5	60 ± 2	65 ± 1	63 ± 1
Methane yield	m^3 CH$_4$ t^{-1} waste	8.8 ± 2.6	12.6 ± 1.0	8.1 ± 2.0	10.9 ± 3.0	8.7 ± 1.2	12.9 ± 2.8
Average productivity	m^3 CH$_4$ m^{-3}·day^{-1}	0.25 ± 0.04	0.33 ± 0.07	0.22 ± 0.04	0.26 ± 0.07	0.25 ± 0.04	0.38 ± 0.07

Mean value ± standard deviation.

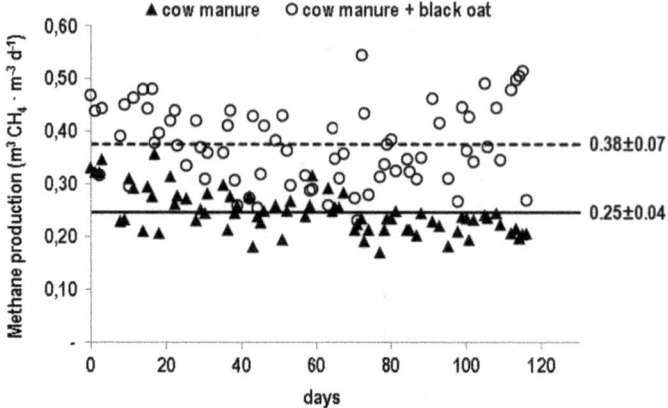

Figure 11 Evolution of volumetric methane production, expressed as m^3 CH_4 m^{-3} day^{-1} at standard temperature and pressure in both the control and the co-digestion reactors (black oat as a co-substrate). Numbers and lines represent mean values ± standard deviations. Source: Riau et al. (2021).

Hence, anaerobic co-digestion with livestock manure was carried out using the ensiled CCs as co-substrates and was compared with a single anaerobic digestion of manure as a control process. Table 6 summarizes the main control parameters of the anaerobic co-digestion process for the three CCs tested. As it can be seen, using CCs as co-substrates increased the organic content in the influent and therefore, the organic loading rate (ORL) rose to between 17% and 46%, while the hydraulic retention time remained constant at 40 days. Generally, the overall process performance was enhanced; for ryegrass, methane yield increased by 43% when increasing the organic load rate by 33%, in comparison with the control digester, whereas organic matter degradation was improved by 16% in terms of volatile solid (VS) removal. Similarly, co-digestion using forage rape as a co-substrate showed a 35% increase of methane yield and the organic matter degradation increased by 15% in terms of VS added. However, it must be taken into account that OLR only increased by 17% in this case. With regards to co-digestion when using black oat, methane yield was increased by 48% while VS removal was only 5%. In this case, the OLR increased significantly (46%) in comparison with the other experiments.

As shown in Fig. 11, using black oat, for instance, the volumetric methane production over time was always higher than in the control digester. Calculations in terms of electric self-sufficiency made on the biogas plant, which provided the liquid fraction for fertilization and the inoculum for the anaerobic digestion tests, foresee an increase in energy production by 42% (from 3.3 to 4.7 mill kWh/a) when CCs are introduced as co-substrates, in comparison to the current plant operation.

10   Conclusion and future trends

The recovery of biobased fertilizers and bioenergy from animal manure to partially replace synthetic mineral fertilizers and fossil fuels should be considered as a key strategy to move towards more sustainable and resilient agriculture and livestock production. This has driven the development of numerous technologies over the last few decades. Some technologies have already been consolidated and widely implemented, such as anaerobic digestion or composting, for bioenergy and biofertilizer production, respectively, but they can still be optimized in a particular context and included as relevant technologies in a biorefinery approach. Other well-known technologies, like solid/liquid separation, maximize nutrient recovery and improve further treatment of the derived fractions, but its efficiency is highly variable and dependent on many factors.

Treatment of the liquid fraction to produce biobased inorganic fertilizers, such as struvite or ammonium sulphate, have gained interest in recent years and innovative technologies have been developed that are able to generate products with similar characteristics to those of mineral fertilizers. As a source of bioenergy, apart from biogas production, thermochemical manure treatments such as gasification to produce syngas, combustion to generate heat and electricity, or hydrothermal liquefaction/carbonization that can directly convert wet manure into bio-oil or hydrochar, have been developed to obtain manure-based bioenergy. However, the operation on an industrial scale is still quite limited and further optimization studies should be performed to overcome the different limiting factors hampering the application of these innovative technologies.

The technologies for nutrient and bioenergy recovery from manure should not be considered as single treatments but as a combination of processes, ideally in a biorefinery framework and with a circular approach, to close nutrient and energy cycling as much as possible along the manure treatment chain.

11   Where to look for further information

11.1 Further reading

- Nutrient Recovery and Reuse (NRR) in European agriculture. A review of the issues, opportunities, and actions (2016). RISE Foundation. https://risefoundation.eu/wp-content/uploads/2020/07/2016_RISE_NRR_Exec _Summary_EN.pdf.
- Schoumans, O. F., Bouraoui, F., Kabbe, C., Oenema, O. and van Dijk, K. C. Phosphorus management in Europe in a changing world. Ambio 2015. https://doi.org/10.1007/s13280-014-0613-9.

11.2 Websites

- European Biogas Association. https://www.europeanbiogas.eu/.
- Eurostat. https://ec.europa.eu/eurostat.
- EIP-Agri. https://ec.europa.eu/eip/agriculture/en/node.
- EIP-AGRI. Focus group on Circular Horticulture https://eu-cap-network.ec .europa.eu/sites/default/files/publication/2023-05/eip-agri_fg_circular _horticulture_final_report_2019_en.pdf.

11.3 Major international research projects

- Circular Agronomics. https://www.circularagronomics.eu/.
- Nutri2cycle. https://www.nutri2cycle.eu/.
- Fertimanure. https://www.fertimanure.eu/.
- Lex4bio. https://www.lex4bio.eu/.
- LIFE Agriclose: https://agriclose.eu/en/.
- Organic Plus: https://organic-plus.net/.

12   References

Aboudi, K., Álvarez-Gallego, C. J. and Romero-García, L. I. (2016). Biomethanization of sugar beet byproduct by semi-continuous single digestion and co-digestion with cow manure. *Bioresour. Technol.* 200, 311–319. https://doi.org/10.1016/j.biortech .2015.10.051.

Awasthi, S. K., Kumar, M., Sarsaiya, S., Ahluwalia, V., Chen, H., Kaur, G., Sirohi, R., Sindhu, R., Binod, P., Pandey, A., Rathour, R., Kumar, S., Singh, L., Zhang, Z., Taherzadeh, M. J. and Awasthi, M. K. (2022). Multi-criteria research lines on livestock manure biorefinery development towards a circular economy: from the perspective of a life cycle assessment and business models strategies. *J. Cleaner Prod.* 341. https://doi.org/10 .1016/j.jclepro.2022.130862.

Blandin, G., Vervoort, H., Le-Clech, P. and Verliefde, A. R. D. (2016). Fouling and cleaning of high permeability forward osmosis membranes. *J. Water Process Eng.* 9, 161–169. https://doi.org/10.1016/J.JWPE.2015.12.007.

Beggio, G., Peng, W., Lü, F., Cerasaro, A., Bonato, T. and Pivato, A. (2022). Chemically enhanced solid–liquid separation of digestate: suspended solids removal and effects on environmental quality of separated fractions. *Waste Biomass Valor.* 13(2), 1029–1041. https://doi.org/10.1007/s12649-021-01591-y.

Bonmatí-Blasi, A., Cerrillo-Moreno, M. and Riau-Arenas, V. (2020). Systems based on physical-chemical processes: nutrient recovery for cycle closure. In: I. Management Association (Ed.). *Waste Management: Concepts, Methodologies, Tools, and Applications* (pp. 526–558). IGI Global. https://doi.org/10.4018/978-1-7998-1210-4 .ch026.

Brienza, C., Sigurnjak, I., Meier, T., Michels, E., Adani, F., Schoumans, O., Vaneeckhaute, C. and Meers, E. (2021). Techno-economic assessment at full scale of a biogas refinery plant receiving nitrogen rich feedstock and producing renewable energy and biobased fertilizers. *J. Cleaner Prod.* 308, 127408. https://doi.org/10.1016/j.jclepro .2021.127408.

Brown, K., Harrison, J. and Bowers, K. (2018). Struvite precipitation from anaerobically digested dairy manure. *Water Air Soil Pollut.* 229(7). https://doi.org/10.1007/s11270 -018-3855-5.

Budzianowski, W. M. (2016). A review of potential innovations for production, conditioning and utilization of biogas with multiple-criteria assessment. *Renew. Sustain. Energy Rev.* 54, 1148–1171. https://doi.org/10.1016/j.rser.2015.10.054.

Bumharter, C., Bolonio, D., Amez, I., García Martínez, M. J. and Ortega, M. F. (2023). New opportunities for the European Biogas industry: a review on current installation development, production potentials and yield improvements for manure and agricultural waste mixtures. *J. Cleaner Prod.* 388, 135867. https://doi.org/10.1016 /j.jclepro.2023.135867.

Cáceres, R., Flotats, X. and Marfà, O. (2006). Changes in the chemical and physicochemical properties of the solid fraction of cattle slurry during composting using different aeration strategies. *Waste Manage.* 26(10), 1081–1091. https://doi.org/10.1016/j .wasman.2005.06.013.

Cáceres, R., Coromina, N., Malińska, K. and Marfà, O. (2015a). Evolution of process control parameters during extended co-composting of green waste and solid fraction of cattle slurry to obtain growing media. *Bioresour. Technol.* 179, 398–406. https://doi .org/10.1016/j.biortech.2014.12.051.

Cáceres, R., Cunill, C. and Marfà, O. (2015b). Composts que es produeixen a Catalunya: caracterització I viabilitat del seu ús com a substrat. *Quad. AGRARIS ICEA* 39, 7–34.

Cáceres, R., Magri, A. and Marfà, O. (2015c). Nitrification of leachates from manure composting under field conditions and their use in horticulture. *Waste Manage.* 44, 72–81. http://doi.org/10.1016/j.wasman.2015.07.039.

Cáceres, R., Malińska, K. and Marfà, O. (2018). Nitrification within composting: a review. *Waste Manage.* 72, 119–137. https://doi.org/10.1016/j.wasman.2017.10.049.

Cáceres, R., Biel, C. and Ortiz, C. (2022). Composting as a nature-based solution for coping with circularity challenges in horticulture: the use of nitrified compost in sustainable horticulture. *Acta Hortic.* (1355), 149–156. https://doi.org/10.17660/ ActaHortic.2022.1355.19.

Cáceres, R. and Parera, J. (2022). Guia per al compostatge en granja de dejeccions ramaderes. Manual for the on-farm composting of livestock droppings. *Departament d'Acció Climàtica*, Alimentació i Agenda Rural (DACC). Barcelona. , 108 pages. Available (free (catalán language)) in the Website at: https://agriclose.eu/wp -content/uploads/2023/04/Guia_compostatge_AGRICLOSE-1.pdf

Camilleri-Rumbau, M. S., Briceño, K., Fjerbæk Søtoft, L., Christensen, K. V., Roda-Serrat, M. C., Errico, M. and Norddahl, B. (2021). Treatment of manure and digestate liquid fractions using membranes: opportunities and challenges. *Int. J. Environ. Res. Public Health* 18(6), 3107. https://doi.org/10.3390/ijerph18063107.

Carreras-Sempere, M., Caceres, R., Viñas, M. and Biel, C. (2021). Use of recovered struvite and ammonium nitrate in fertigation in tomato (Lycopersicum esculentum) production for boosting circular and sustainable horticulture. *Agriculture* 11(11). https://doi.org/10.3390/agriculture11111063.

Carreras-Sempere, M., Biel, C., Viñas, M., Guivernau, M. and Caceres, R. (2022). The use of recovered struvite and ammonium nitrate in fertigation in a Horticultural rotation: agronomic and microbiological assessment. *Environ.Technol. (UK)*. https://doi .org/10.1080/09593330.2022.2154172.

Cerrillo, M., Palatsi, J., Comas, J., Vicens, J. and Bonmatí, A. (2015). Struvite precipitation as a technology to be integrated in a manure anaerobic digestion treatment plant - removal efficiency, crystal characterization and agricultural assessment. *J. Chem. Technol. Biotechnol.* 90(6), 1135–1143. https://doi.org/10.1002/jctb.4459.

Cerrillo, M., Viñas, M. and Bonmatí, A. (2016a). Removal of volatile fatty acids and ammonia recovery from unstable anaerobic digesters with a microbial electrolysis cell. *Bioresour. Technol.* 219, 348–356. https://doi.org/10.1016/j.biortech.2016.07.103.

Cerrillo, M., Viñas, M. and Bonmatí, A. (2016b). Overcoming organic and nitrogen overload in thermophilic anaerobic digestion of pig slurry by coupling a microbial electrolysis cell. *Bioresour. Technol.* 216, 362–372. https://doi.org/10.1016/j.biortech.2016.05.085.

Cerrillo, M., Burgos, L. and Bonmatí, A. (2021a). Biogas upgrading and ammonia recovery from livestock manure digestates in a combined electromethanogenic biocathode–hydrophobic membrane system. *Energies* 14(2). https://doi.org/10.3390/en14020503.

Cerrillo, M., Burgos, L., Noguerol, J., Riau, V. and Bonmatí, A. (2021b). Ammonium and phosphate recovery in a three chambered microbial electrolysis cell: towards obtaining struvite from livestock manure. *Processes* 9(11), 1916. https://doi.org/10.3390/pr9111916.

Cerrillo, M., Riau, V. and Bonmatí, A. (2023). Recent advances in bioelectrochemical systems for nitrogen and phosphorus recovery using membranes. *Membranes* 13(2), 186. https://doi.org/10.3390/membranes13020186.

Cheng, S. and Logan, B. E. (2011). High hydrogen production rate of microbial electrolysis cell (MEC) with reduced electrode spacing. *Bioresour. Technol.* 102(3), 3571–3574. https://doi.org/10.1016/j.biortech.2010.10.025.

Cocolo, G., Curnis, S., Hjorth, M. and Provolo, G. (2012). Effect of different technologies and animal manures on solid-liquid separation efficiencies. *J. Agric. Eng.* 43(2). https://doi.org/10.4081/jae.2012.e9.

Corona, F., Hidalgo, D., Martín-Marroquín, J. M. and Meers, E. (2021). Study of pig manure digestate pre-treatment for subsequent valorisation by struvite. *Environ. Sci. Pollut. Res. Int.* 28(19), 24731–24743. https://doi.org/10.1007/s11356-020-10918-6.

Cusick, R. D., Ullery, M. L., Dempsey, B. A. and Logan, B. E. (2014). Electrochemical struvite precipitation from digestate with a fluidized bed cathode microbial electrolysis cell. *Water Res.* 54, 297–306. Available at: https://doi.org/10.1016/j.watres.2014.01.051.

Czekala, W., Malinska, K., Cáceres, R., Janczak, D., Dach, J. and Lewicki, A. (2016). Co-composting of poultry manure mixtures amended with biochar - the effect of biochar on temperature and C-CO2 emission. *Bioresour. Technol.* 200, 921–927. https://doi.org/10.1016/j.biortech.2015.11.019.

Directive of the Council of 12 December 1991 Concerning the Protection of Waters against Pollution Caused by Nitrates from Agricultural Sources (91/676/EC).

Ehmann, A., Thumm, U. and Lewandowski, I. (2018). Fertilizing potential of separated biogas digestates in annual and perennial biomass production systems. *Front. Sustain. Food Syst.* 2. https://doi.org/10.3389/fsufs.2018.00012.

Eurostat (2022). Agri-Environmental Indicator - Mineral Fertilizer Consumption - Statistics Explained. Available at: europa.eu.

Fangueiro, D., Coutinhob, J., Borges, L. and Vasconcelos, E. (2015). Recovery efficiency of nitrogen from liquid and solid fractions of pig slurry obtained using different separation technologies. *J. Plant Nutr. Soil Sci.* 178(2), 229–236. https://doi.org/10 .1002/jpln.201400261.

Flotats Ripoll, X., Foged, H., Bonmatí Blasi, A., Palatsi Civit, J., Magrí Aloy, A. and Schelde, K. M. (2011). Manure processing technologies. Technical Report No. II concerning. Manure Processing Activities in Europe to the European Commission, Directorate-General Environment. Project reference. EnvironmentalistB.1/ETU/2010/0007, 184 pp. Available at: http://hdl.handle.net/2117/18944.

Fukumoto, Y. and Haga, K. (2004). Advanced treatment of swine wastewater by electrodialysis with a tubular ion exchange membrane. *Anim. Sci. J.* 75(5), 479–485. https://doi.org/10.1111/j.1740-0929.2004.00216.x.

Guo, X. and Jin, X. (2014). Purification of UF-treated anaerobically digested manure wastewater by two-pass reverse osmosis. *Desalin. Water Treat.* 52(16–18), 3027–3034. https://doi.org/10.1080/19443994.2013.797671.

Haddadi, S., Nabi-Bidhendi, G. and Mehrdadi, N. (2014). Nitrogen removal from wastewater through microbial electrolysis cells and cation exchange membrane. *J. Environ. Health Sci. Eng.* 12(1), 48. https://doi.org/10.1186/2052-336X-12-48.

Haug, R. T. (1993). *The Practical Handbook of Compost Engineering.* ISBN: 0-87371-373-7. 715 pp. https://doi.org/10.1201/9780203736234.

Hendriks, C. M. J., Shrivastava, V., Sigurnjak, I., Lesschen, J. P., Meers, E., Noort, Rv., Yang, Z. and Rietra, R. P. J. J. (2022). Replacing mineral fertilizers for bio-based fertilizers in potato growing on sandy soil: A case study. *Appl. Sci.* 12(1), 341. https://doi.org/10 .3390/app12010341.

Horváth, I. S., Tabatabaei, M., Karimi, K. and Kumar, R. (2016). Recent updates on biogas production - a review. *Biofuel Res. J.* 3(2), 394–402. https://doi.org/10.18331/ BRJ2016.3.2.4.

Huang, H., Xu, C. and Zhang, W. (2011). Removal of nutrients from piggery wastewater using struvite precipitation and pyrogenation technology. *Bioresour. Technol.* 102(3), 2523–2528. https://doi.org/10.1016/j.biortech.2010.11.054.

Janczak, D., Malińska, K., Czekała, W., Cáceres, R., Lewicki, A. and Dach, J. (2017). Biochar to reduce ammonia emissions in gaseous and liquid phase during composting of poultry manure with wheat straw. *Waste Manage.* 66, 36–45. https://doi.org/10.1016 /j.wasman.2017.04.033.

Jorgensen, K. and Jensen, L. S. (2009). Chemical and biochemical variation in animal manure solids separated using different commercial separation technologies. *Bioresour. Technol.* 100(12), 3088–3096. https://doi.org/10.1016/j.biortech.2009.01 .065.

Khoshnevisan, B., Duan, N., Tsapekos, P., Awasthi, M. K., Liu, Z., Mohammadi, A., Angelidaki, I., Tsang, D. C., Zhang, Z., Pan, J., Ma, L., Aghbashlo, M., Tabatabaei, M. and Liu, H. (2021). A critical review on livestock manure biorefinery technologies: sustainability, challenges, and future perspectives. *Renew. Sustain. Energy Rev.* 135, 110033. https://doi.org/10.1016/j.rser.2020.110033.

Koffi, N. J. and Okabe, S. (2020). High voltage generation from wastewater by microbial fuel cells equipped with a newly designed low voltage booster multiplier (LVBM). *Sci. Rep.* 10(1), 18985. https://doi.org/10.1038/s41598-020-75916-7.

Konieczny, K., Kwiecińska, A. and Gworek, B. (2011). The recovery of water from slurry produced in high density livestock farming with the use of membrane processes. *Sep. Purif. Technol.* 80(3), 490–498. https://doi.org/10.1016/j.seppur.2011.06.002.

Kovačić, Đ., Lončarić, Z., Jović, J., Samac, D., Popović, B. and Tišma, M. (2022). Digestate management and processing practices: a review. *Appl. Sci.* 12(18), 9216. https://doi .org/10.3390/app12189216.

Lorick, D., Macura, B., Ahlström, M., Grimvall, A. and Harder, R. (2020). Effectiveness of struvite precipitation and ammonia stripping for recovery of phosphorus and nitrogen from anaerobic digestate: a systematic review. *Environ. Evid.* 9(1), 27. https://doi.org/10.1186/s13750-020-00211-x.

Lyons, G. A., Cathcart, A., Frost, J. P., Wills, M., Johnston, C., Ramsey, R. and Smyth, B. (2021). Review of two mechanical separation technologies for the sustainable management of agricultural phosphorus in nutrient-vulnerable zones. *Agronomy* 11(5), 836. https://doi.org/10.3390/agronomy11050836.

Masse, L., Massé, D. I. and Pellerinb, Y. (2007). The use of membranes for the treatment of manure: a critical literature review. *Biosyst. Eng.* 98(4), 371–380. https://doi.org/10 .1016/J.BIOSYSTEMSENG.2007.09.003.

Møller, H. B., Lund, I. and Sommer, S. G. (2000). Solid-liquid separation of livestock slurry: efficiency and cost. *Bioresour. Technol.* 74, 223–229. https://doi.org/10.1016/S0960 -8524%2800%2900016-X.

Møller, H. B., Sommer, S. G. and Ahring, B. K. (2002). Separation efficiency and particle size distribution in relation to manure type and storage conditions. *Bioresour. Technol.* 85(2), 189–196. https://doi.org/10.1016/s0960-8524(02)00047-0.

Nagarajan, A., Goyette, B., Raghavan, V., Bhaskar, A. and Rajagopal, R. (2023). Nutrient recovery via struvite production from livestock manure-digestate streams: towards closed loop bioeconomy. *Process Saf. Environ. Prot.* 171, 273–288. https://doi.org /10.1016/j.psep.2023.01.006.

Nkoa, R. (2014). Agricultural benefits and environmental risks of soil fertilization with anaerobic digestates: a review. *Agron. Sustain. Dev.* 34(2), 473–492. https://doi.org /10.1007/s13593-013-0196-z.

Pikaar, I., Guest, J., Ganigué, R., Jensen, P., Rabaey, K., Seviour, T., Trimmer, J., van der Kolk, O., Vaneeckhaute, C. and Verstraete, W. (2022). Resource recovery from water: principles and application. In: IWA Publishing (Ed.). ISBN electronic: 9781780409566. https://doi.org/10.2166/9781780409566.

Pinnekamp, J. and Friedrich, H. (2006). Membrane technology for waste water treatment. *Municipal Water Waste Management.* FiW-Verlag. ISBN 978-3-939377-01-6 2.

Prado, J., Ribeiro, H., Alvarenga, P. and Fangueiro, D. (2022). A step towards the production of manure-based fertilizers: disclosing the effects of animal species and slurry treatment on their nutrients content and availability. *J. Cleaner Prod.* 337. https://doi .org/10.1016/j.jclepro.2022.130369.

Prenafeta-Boldú, F. X. i Parera, J. (2020). Guia de les tecnologies de tractament de les dejeccions ramaderes a Catalunya, Pesca i Alimentació (DARP), Barcelona, 72 pàgines. *Departament d'Agricultura, Ramaderia.* Available at: https://www .agrodigital.com/wp-content/uploads/2020/12/Guiadeyeccionesc.pdf.

Regueiro, I., Coutinho, J., Balsari, P., Popovic, O. and Fangueiro, D. (2016). Acidification of pig slurry before separation to improve slurry management on farms. *Environ. Technol.* 37(15), 1906–1913. https://doi.org/10.1080/09593330.2015.1135992.

Regulation (EU) 2019/1009 of the European Parliament and of the Council of 5 June 2019 laying down rules on the making available on the market of EU fertilising products and amending Regulations (EC) 1069/2009) and (EC) No 1107/2009 and repealing

Regulation (EC)(2003/2003). Available at: https://eur-lex.europa.eu/eli/reg/2019 /1009/oj

Riau, V., Burgos, L., Camps, F., Domingo, F., Torrellas, M., Antón, A. and Bonmatí, A. (2021). Closing nutrient loops in a maize rotation. Catch crops to reduce nutrient leaching and increase biogas production by anaerobic co-digestion with dairy manure. *Waste Manage.* 126, 719–727. https://doi.org/10.1016/j.wasman.2021.04.006.

Rizzioli, F., Bertasini, D., Bolzonella, D., Frison, N. and Battista, F. (2023). A critical review on the techno-economic feasibility of nutrients recovery from anaerobic digestate in the agricultural sector. *Sep. Purif. Technol.* 306, 122690. https://doi.org/10.1016/j .seppur.2022.122690.

Robles-Aguilar, A. A., Pang, J., Postma, J. A., Schrey, S. D., Lambers, H. and Jablonowski, N. D. (2019). The effect of pH on morphological and physiological root traits of *Lupinus angustifolius* treated with struvite as a recycled phosphorus source. *Plant Soil* 434(1- 2), 65–78. https://doi.org/10.1007/s11104-018-3787-2.

Robles-Aguilar, A. A., Schrey, S. D., Postma, J. A., Temperton, V. M. and Jablonowski, N. D. (2020). Phosphorus uptake from struvite is modulated by the nitrogen form applied. *J. Plant Nutr. Soil Sci.* 183(1), 80–90. https://doi.org/10.1002/jpln.201900109.

Rodríguez Arredondo, M., Kuntke, P., ter Heijne, A. and Buisman, C. J. (2019). The concept of load ratio applied to bioelectrochemical systems for ammonia recovery. *J. Chem. Technol. Biotechnol.* 94(6), 2055–2061. Available at: https://doi.org/10.1002/jctb .5992.

Sawatdeenarunat, C., Nguyen, D., Surendra, K. C., Shrestha, S., Rajendran, K., Oechsner, H., Xie, L. and Khanal, S. K. (2016). Anaerobic biorefinery: current status, challenges, and opportunities. *Bioresour. Technol.* 215, 304–313. https://doi.org/10.1016/j .biortech.2016.03.074.

Sena, M. and Hicks, A. L. (2018). Life cycle assessment review of struvite precipitation in wastewater treatment. *Resour. Conserv. Recycl.* 139, 194–204. https://doi.org/10 .1016/J.RESCONREC.2018.08.009.

Shi, L., Simplicio, W. S., Wu, G., Hu, Z., Hu, H. and Zhan, X. (2018). Nutrient recovery from digestate of anaerobic digestion of livestock manure: a review. *Curr. Pollut. Rep.* 4(2), 74–83. https://doi.org/10.1007/s40726-018-0082-z.

Shi, S., Tong, B., Wang, X., Luo, W., Tan, M., Wang, H. and Hou, Y. (2022). Recovery of nitrogen and phosphorus from livestock slurry with treatment technologies: A meta-analysis. *Waste Manage.* 144, 313–323. https://doi.org/10.1016/j.wasman.2022.03.027.

Słupek, E., Makoś, P. and Gębicki, J. (2020). Theoretical and economic evaluation of low-cost deep eutectic solvents for effective biogas upgrading to bio-methane. *Energies* 13(13), 3379. https://doi.org/10.3390/en13133379.

Song, Y.-H., Hidayat, S., Effendi, A. J. and Park, J. (2021). Simultaneous hydrogen production and struvite recovery within a microbial reverse-electrodialysis electrolysis cell. *J. Ind. Eng. Chem.* 94, 302–308. https://doi.org/10.1016/j.jiec.2020.10.043.

Sotres, A., Cerrillo, M., Viñas, M. and Bonmatí, A. (2015). Nitrogen recovery from pig slurry in a two-chambered bioelectrochemical system. *Bioresour. Technol.* 194, 373–382. https://doi.org/10.1016/j.biortech.2015.07.036.

Takkellapati, S., Li, T. and Gonzalez, M. A. (2018). An overview of biorefinery-derived platform chemicals from a cellulose and hemicellulose biorefinery. *Clean Technol. Environ. Policy* 20(7), 1615–1630. https://doi.org/10.1007/s10098-018-1568-5.

Tao, W., Fattah, K. P. and Huchzermeier, M. P. (2016). Struvite recovery from anaerobically digested dairy manure: a review of application potential and hindrances. *J. Environ. Manage.* 169, 46–57. https://doi.org/10.1016/j.jenvman.2015.12.006.

Tian, P., Gong, B., Bi, K., Liu, Y., Ma, J., Wang, X., Ouyang, Z. and Cui, X. (2023). Anaerobic co-digestion of pig manure and rice straw: optimization of process parameters for enhancing biogas production and system stability. *Int. J. Environ. Res. Public Health* 20(1), 804. https://doi.org/10.3390/ijerph20010804.

Torrellas, M., Burgos, L., Tey, L., Noguerol, J., Riau, V., Palatsi, J., Antón, A., Flotats, X. and Bonmatí, A. (2018). Different approaches to assess the environmental performance of a cow manure biogas plant. *Atmos. Environ.* 177, 203-213. https://doi.org/10.1016/j.atmosenv.2018.01.023.

Tur-Cardona, J., Bonnichsen, O., Speelman, S., Verspecht, A., Carpentier, L., Debruyne, L., Marchand, F., Jacobsen, B. H. and Buysse, J. (2018). Farmers' reasons to accept bio-based fertilizers: A choice experiment in seven different European countries. *J. Cleaner Prod.* 197, 406-416. https://doi.org/10.1016/j.jclepro.2018.06.172.

Varol, A. and Ugurlu, A. (2017). Comparative evaluation of biogas production from dairy manure and co-digestion with maize silage by CSTR and new anaerobic hybrid reactor. *Eng. Life Sci.* 17(4), 402-412. https://doi.org/10.1002/elsc.201500187.

Velthof, G. L. (2011). *Synthesis of the Research within the Framework of the Mineral Concentrates Pilot*. Available at: https://edepot.wur.nl/192069. 72 pp. Wageningen, Alterra. Alterra report 2224.

Villano, M., Monaco, G., Aulenta, F. and Majone, M. (2011). Electrochemically assisted methane production in a biofilm reactor. *J. Power Sources* 196(22), 9467-9472. https://doi.org/10.1016/j.jpowsour.2011.07.016.

Wagner, E. and Karthikeyan, K. G. (2022). Precipitating phosphorus as struvite from anaerobically-digested dairy manure. *J. Cleaner Prod.* 339, 130675. https://doi.org/10.1016/j.jclepro.2022.130675.

Yetilmezsoy, K. and Sapci-Zengin, Z. (2009). Recovery of ammonium nitrogen from the effluent of UASB treating poultry manure wastewater by MAP precipitation as a slow release fertilizer. *J. Hazard. Mater.* 166(1), 260-269. https://doi.org/10.1016/j.jhazmat.2008.11.025.

Zeppilli, M., Mattia, A., Villano, M. and Majone, M. (2017). Three-chamber bioelectrochemical system for biogas upgrading and nutrient recovery. *Fuel Cells* 17(5), 593-600. https://doi.org/10.1002/fuce.201700048.

Zeppilli, M., Simoni, M., Paiano, P. and Majone, M. (2019). Two-side cathode microbial electrolysis cell for nutrients recovery and biogas upgrading. *Chem. Eng. J.* 370, 466-476. https://doi.org/10.1016/j.cej.2019.03.119.

Zeppilli, M., Cristiani, L., Dell'Armi, E. and Majone, M. (2020). Bioelectromethanogenesis reaction in a tubular microbial electrolysis cell (MEC) for biogas upgrading. *Renew. Energy* 158, 23-31. Available at: https://doi.org/10.1016/j.renene.2020.05.122.

Zhang, Y. and Angelidaki, I. (2015). Counteracting ammonia inhibition during anaerobic digestion by recovery using submersible microbial desalination cell. *Biotechnol. Bioeng.* 112(7), 1478-1482. https://doi.org/10.1002/bit.25549.

Zhang, Z., Xu, Z., Song, X., Zhang, B., Li, G., Huda, N. and Luo, W. (2020). Membrane processes for resource recovery from anaerobically digested livestock manure effluent: opportunities and challenges. *Curr. Pollut. Rep.* 6(2), 123-136. https://doi.org/10.1007/s40726-020-00143-7.

Zhu, Q. L., Wu, B., Pisutpaisal, N., Wang, Y. W., Ma, K. D., Dai, L. C., Qin, H., Tan, F. R., Maeda, T., Xu, Y. S., Hu, G. Q. and He, M. X. (2021). Bioenergy from dairy manure: technologies, challenges and opportunities. *Sci. Total Environ.* 790, 148199. https://doi.org/10.1016/j.scitotenv.2021.148199.